Global Warming Handbook

Global Warming Handbook

Edited by Sharon Manuel

MURPHY & MOORE
www.murphy-moorepublishing.com

Published by Murphy & Moore Publishing,
1 Rockefeller Plaza,
New York City, NY 10020, USA
www.murphy-moorepublishing.com

Global Warming Handbook
Edited by Sharon Manuel

International Standard Book Number: 978-1-63987-261-9 (Hardback)

Cataloging-in-Publication Data

Global warming handbook / edited by Sharon Manuel.
    p. cm.
Includes bibliographical references and index.
ISBN 978-1-63987-261-9
1. Global warming. 2. Global temperature changes. 3. Greenhouse effect, Atmospheric.
4. Climatic changes. 5. Climatology. I. Manuel, Sharon.
QC981.8.G56 G56 2022
363.738 74--dc23

TABLE OF CONTENTS

PREFACE

This book aims to highlight the current researches and provides a platform to further the scope of innovations in this area. This book is a product of the combined efforts of many researchers and scientists, after going through thorough studies and analysis from different parts of the world. The objective of this book is to provide the readers with the latest information of the field.

Global warming refers to the phenomenon of increase in the average air temperatures near the surface of the Earth over the past few centuries. Human activities since the beginning of the Industrial Revolution are the driving factors behind the accelerated climate change. Activities such as deforestation and burning of fossil fuel lead to an increase in the concentration of greenhouse gases which in turn leads to heat getting trapped in the atmosphere. The rise of the global average temperature has resulted in significant societal, economic and ecological damage. Continued rise in temperature can lead to increased extinction of many plant and animal species, shifts in agricultural patterns and rising sea levels. This book explores all the important aspects of global warming in the present day scenario. Different approaches, evaluations, methodologies and advanced studies on this field have been included herein. This book will help new researchers by foregrounding their knowledge in this branch.

I would like to express my sincere thanks to the authors for their dedicated efforts in the completion of this book. I acknowledge the efforts of the publisher for providing constant support. Lastly, I would like to thank my family for their support in all academic endeavors.

Editor

Influence of Climate Change on Weed Vegetation

Vytautas Pilipavičius

1. Introduction

Climate is the biggest abiotical factor influencing the whole vegetation. At climate changing conditions adaptation ability of vegetation changes to grow in certain territory. Competitive abilities of plants are changing showing through new plant and weed biological qualities.

Global warming and climate change refer to an increase in average global temperatures. Global warming is primarily caused by increases in "greenhouse" gases (GHG). A warming planet thus leads to climate changes which can adversely affect weather in different ways. Some of the prominent indicators for a global warming are: temperature over land, snow cover and glaciers on hills, ocean heat content, sea ice, sea level, sea surface temperature, temperature over ocean, humidity, tropospheric temperature. Global warming in today's scenario is threat to the survival of mankind [55]. Climate change inspired by global warming could lead to change of natural climatic zones, i.e. Tundra would disappear, Taiga would decrease essentially, Mediterranean climate zone would decrease and move to north, deserts and Arid world zones would move 400-800 km north to populous subtropical areas, main agricultural zones would move to north areas with low-fertile and worse soils [56, 57]. Global warming is closely associated as well with a broad spectrum of other climate changes, such as increases in the frequency of intense rainfall, decreases in snow cover and sea ice, more frequent and intense heat waves, rising sea levels, and widespread ocean acidification [55].

"The damage that climate change is causing and that will get worse if we fail to act goes beyond the hundreds of thousands of lives, homes and businesses lost, ecosystems destroyed, species driven to extinction, infrastructure smashed and people inconvenienced." – David Suzuki[1]

---

1 Suzuki D. BrainyQuote.com, XploreInc, 2014. Available from http://www.brainyquote.com/quotes/quotes/d/davidsuzuk471841.html

Seasons of the years are constantly attended by the increase of marginal air conditions. Many researchers agree that anthropogenic activity is reason for climate change and it induced changes of the nature [1]. Agriculture and forestry take important place in Lithuanian national economy, therefore it is actual to adjust those sectors to climate change for mitigating consequences [2]. The constant competition between agricultural plants and weeds is seen in agro-ecosystems when the yields minimize. Alongside with other factors its progress can be explained by the plant resistance to abiotical factors. Different sensitivity of various varieties and their adaptability to the human activity may govern their relationship in agro-ecosystems. Thus, the adaptability of different abiotical factors for both agricultural plants and weeds should be estimated [3]. Weed spreading regularities are significantly dependent on weed ability to adapt, that is to adapt to changeable factors of environment.

Analogous weed chemical composition to agricultural plants induces competition in agro-phytocenoses for growth factors. Weeds occupied place where agricultural plants could grow [4]. Adaptation possibility of separate plant species is different and can vary their competition as environment conditions change. It can cause serious agricultural problems. Undesirable change of plant species follows when environment conditions vary in ecosystems. Usually weeds signify by higher plasticity [5].

Biological invasions and climate warming are two major threats to the world's biodiversity. To date, their impacts have largely been considered independently, despite indications that climate warming may increase the success of many invasive alien species [50].

The climate system is a complex, interactive system consisting of the atmosphere, land surface, snow and ice, oceans and other bodies of water, and living things. Climate is usually described in terms of the mean and variability of temperature, precipitation and wind over a period of time, ranging from months to millions of years (the classical period is 30 years) [6]. Observations of the climate system are based on direct measurements and remote sensing from satellites and other platforms. Global-scale observations from the instrumental era began in the mid-19th century for temperature and other variables. Paleoclimate reconstructions extended some records back hundreds to millions of years [7].

Changes in the atmospheric abundance of greenhouse gases and aerosols, in solar radiation and in land surface properties alter the energy balance of the climate system [8]. Global GHG emissions due to human activities have grown since pre-industrial times, with an increase of 70% between 1970 and 2004 (Figure 1). Annual CO_2 emissions from fossil fuel combustion and cement production were 8.3 GtC12 yr^{-1} averaged over 2002-2011 and were 9.5 GtC yr^{-1} in 2011, 54% more than the level in 1990. Annual net CO_2 emissions from anthropogenic land use change were 0.9 GtC yr^{-1} on average during 2002 to 2011 [7]. The global atmospheric concentration of carbon dioxide has increased from a pre-industrial value of about 280 ppm to 379 ppm in 2005 (Figure 2). The annual carbon dioxide concentration growth rate was larger during the last 10 years (1995-2005 average: 1.9 ppm per year), than it has been during 1960-2005 (average: 1.4 ppm per year) although there is year-to-year variability in growth rates [8]. In 2011 the concentrations of CO_2 were 391 ppm, and exceeded the pre-industrial levels by about 40% [7].

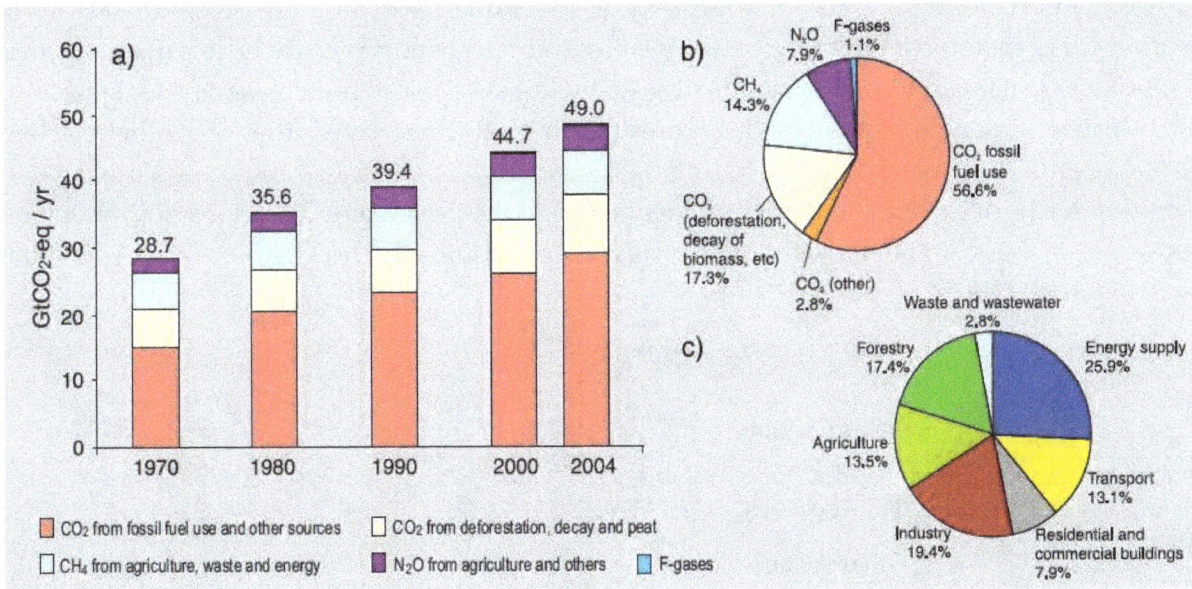

Figure 1. a) Global annual emissions of anthropogenic GHGs from 1970 to 2004. (b) Share of different anthropogenic GHGs in total emissions in 2004 in terms of carbon dioxide equivalents (CO_2-eq). (c) Share of different sectors in total anthropogenic GHG emissions in 2004 in terms of CO_2-eq (Forestry includes deforestation). Source: Climate Change 2007: Synthesis Report [9]

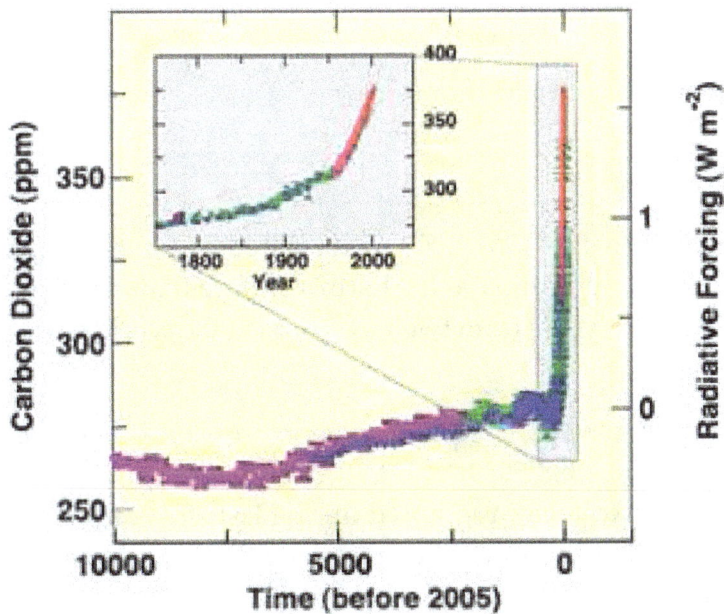

Figure 2. Atmospheric concentrations of carbon dioxide over the last 10,000 years (large panels) and since 1750 (inset panels). Measurements are shown from ice cores (symbols with different colours for different studies) and atmospheric samples (red lines). The corresponding radiative forcings are shown on the right hand axes of the large panels. Source: IPCC, 2007: Summary for Policymakers [8]

The great part of GHG emissions locally, i.e. in Lithuania evaluating separate sectors of economy, is generated from energy supply objects and transport (Figure 3). It is in accordance with other developed industrial countries. As well it is forecasted increase of CO_2 emissions till 2030 in all sectors of economy in Lithuania (Figure 3). Lithuania together with other modern world countries work solving global climate change problems. In Lithuania annual GHG emission terms of carbon dioxide equivalents (CO_2-eq) covered about 4-5 tons per inhabitant and is one of the lowest in European countries where annual GHG of CO_2-eq is about 3-15 tons per inhabitant [10].

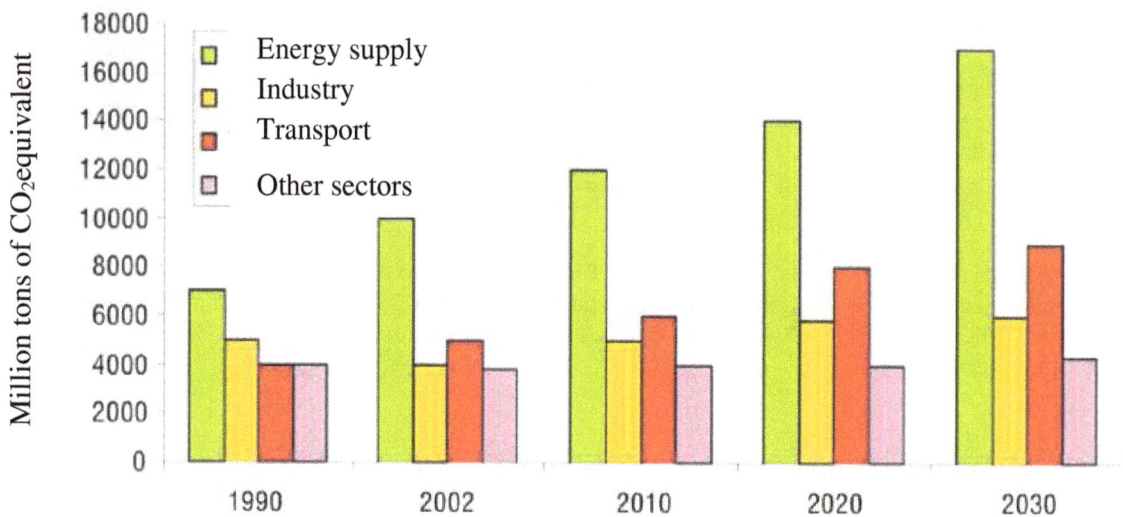

Figure 3. Present and forecasted share of anthropogenic GHGs in total emission in terms of carbon dioxide equivalent (CO_2-eq) in Lithuania. Source: Ministry of Environment of the Republic of Lithuania [10]

Global warming is the increase of the average temperature of the Earth's near-surface air and the oceans since the mid-twentieth century and its projected continuation. Global mean surface temperature anomaly relative to 1961–1990 is presented in figure 4. Each of the last three decades has been successively warmer at the Earth's surface than any preceding decade since 1850 [7]. The warmest eleven years from twelve records in the world since 1850 were stated in the period of 1995-2006 [8].

In Lithuania there was registered unique climate expression – even seven warm winters successively in the period of 1988/1989 – 1994/1995. Such long period of anomalously warm winters in the Baltic region were not registered during last 200 years [13]. Climate changes in Lithuania manifest through increasing air temperatures and precipitation during winter and slightly increase of air temperatures and decrease of precipitation during summer time [14].

Dynamics of average air temperature in Lithuania during 1961–2006 is presented in figure 5. The results from three locations, i.e. Klaipėda, Kaunas and Vilnius, showed increasing calculated trend-lines (dotted lines) and actual air temperature variation (solid lines).

Lithuanian average year air temperature in 1991–2006 increased by 0.7-1.0 °C relatively to 1961-1990 (Figure 6). That shows fast local climate warming in Lithuania. Climate warming

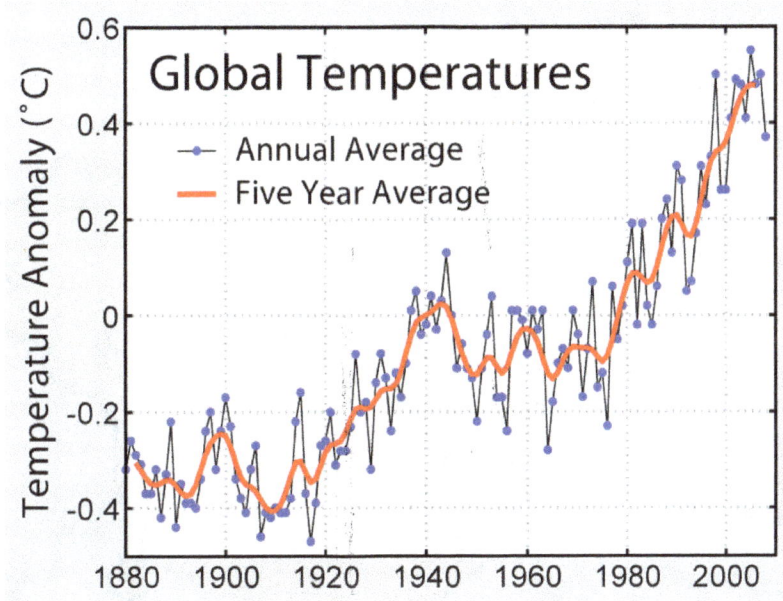

Figure 4. Global mean surface temperature anomaly relative to 1961–1990. Source: Climate Change 2007: The Physical Science Basis [8, 11, 12]

tendencies are the most expressed in North and West Lithuania. Therefore, the last 16 year (1991-2006) average air temperature in Lithuania overcame the limit of 6°C.

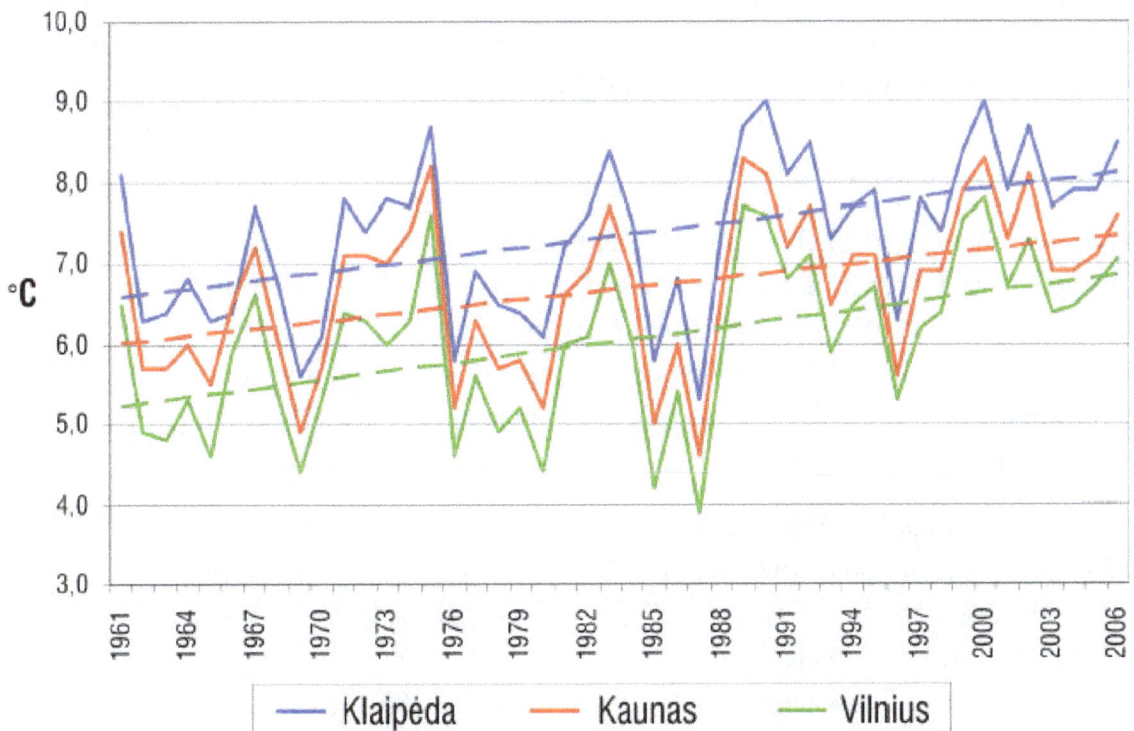

Figure 5. Average year air temperature (°C) change dynamics linear trends (dotted lines) in Lithuania during 1961–2006. Source: Lithuanian Hydrometeorological Service under the Ministry of Environment [10]

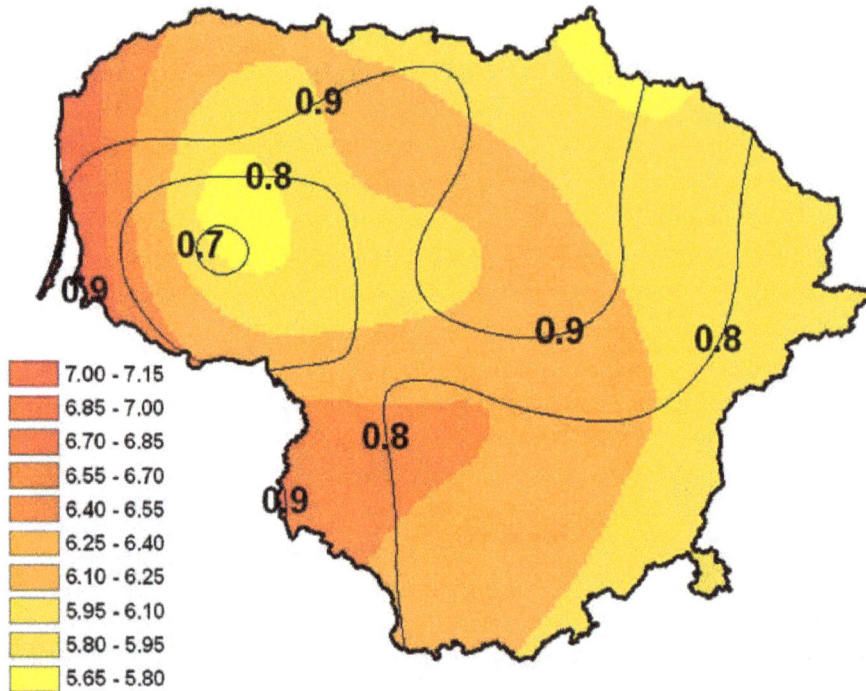

Figure 6. Lithuanian average year air temperature (°C) in 1961–1990. The air temperature differences between 1961-1990 and 1991-2006 are shown by isolines. Source: Lithuanian Hydrometeorological Service under the Ministry of Environment [10]

During the second half of the 20th century and early part of the 21st century, global average surface temperature increased and sea level rose. Over the same period, the amount of snow cover in the Northern Hemisphere decreased (Figure 7). If radiative forcing was to be stabilised in 2100 at A1B levels, thermal expansion alone would lead to 0.3 to 0.8 m of sea level rise by 2300 (relatively to 1980–1999) [8].

The best estimates and likely ranges for global average surface air warming for six SRES emissions marker scenarios are shown in figure 8. Including uncertainties in the future greenhouse gas concentrations and climate sensitivity, the IPCC, scientific intergovernmental body set up by the World Meteorological Organization (WMO) and by the United Nations Environment Programme (UNEP), anticipates a warming of 1.1°C to 6.4°C by the end of the 21st century, relatively to 1980–1999 [8].

The globally averaged combined land and ocean surface temperature data show a warming of 0.85°C, over the period of 1880 to 2012. The total increase between the average of the 1850–1900 and the 2003–2012 periods is 0.78°C. For the longest period when calculation of regional trends is sufficiently complete (1901 to 2012), almost the entire globe has experienced surface warming [7]. If radiative forcing was to be stabilised in 2100 at B1or A1B levels (Figure 8), a further increase in global average temperature of about 0.5°C would be still expected, mostly by 2200. Thermal expansion would continue for many centuries, due to the time required to transport heat into the deep ocean [8].

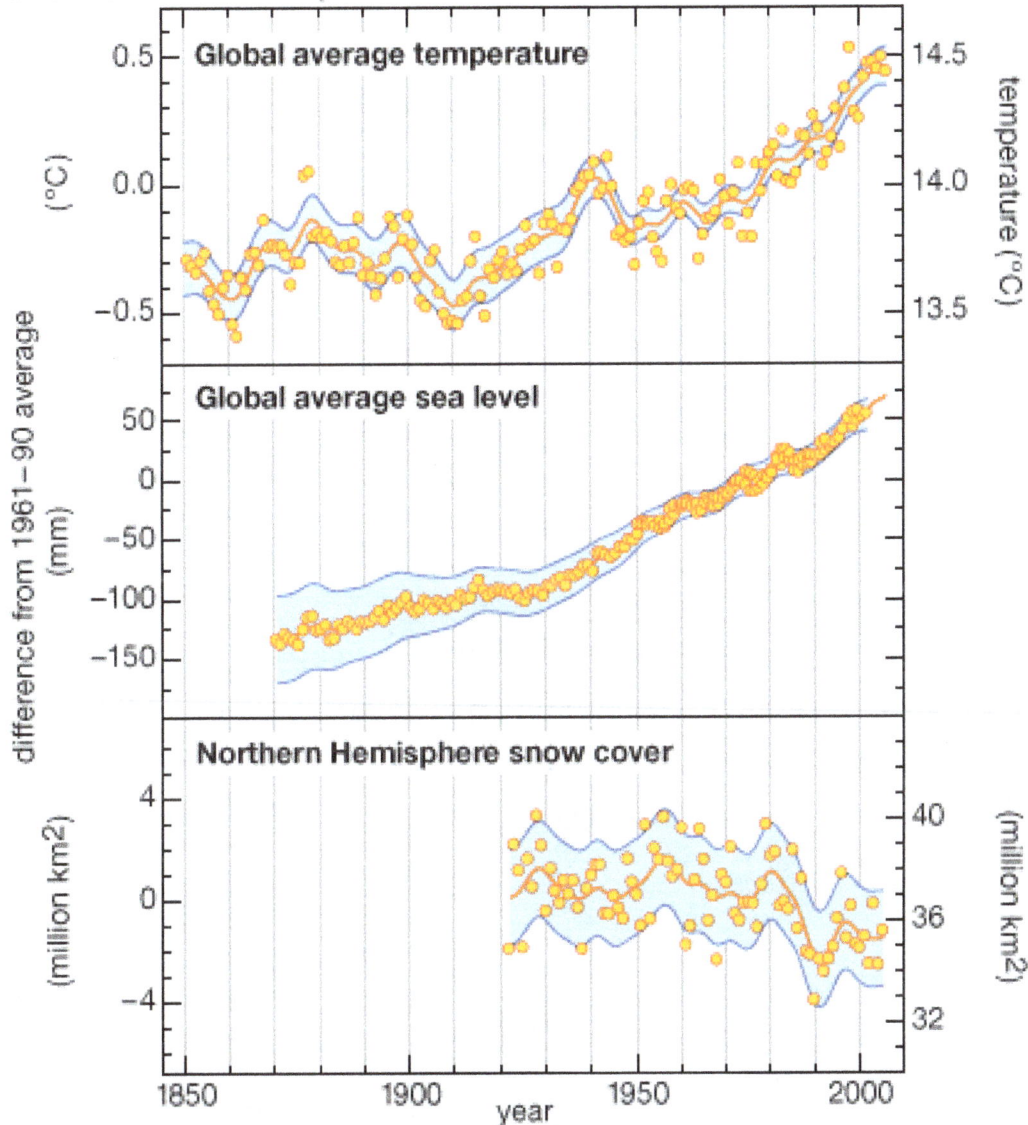

Changes in global average temperature, global average sea level, and Northern Hemisphere snow cover

Source: Climate Change 2007: The Physical Science Basis,
Summary for Policymakers, Intergovernmental Panel on Climate Change

Figure 7. Changes in global average temperature, global average sea level and Northern Hemisphere snow cover. Source: Climate Change 2007: The Physical Science Basis, Summary for Policymakers, IPCC [8]

Over the last two decades, the Greenland and Antarctic ice sheets have been losing mass, glaciers have continued to shrink almost worldwide, and Arctic sea ice and Northern Hemisphere spring snow cover have continued to decrease in extent [7]. Analogous to global tendencies, due to warming climate, day number with snow cover became more unstable and is being decreasing in Lithuania. Day number with snow cover during 1991-2006 relatively to 1961-1990 averagely decreased by 4-10 days (Figure 9). However, during winter emerging maximal snow cover thick increased by 0.8-2.0 cm. It is connected with last year's increase of cold season precipitation amount and more often heavy snowing [10].

Estimations of past and future global warming

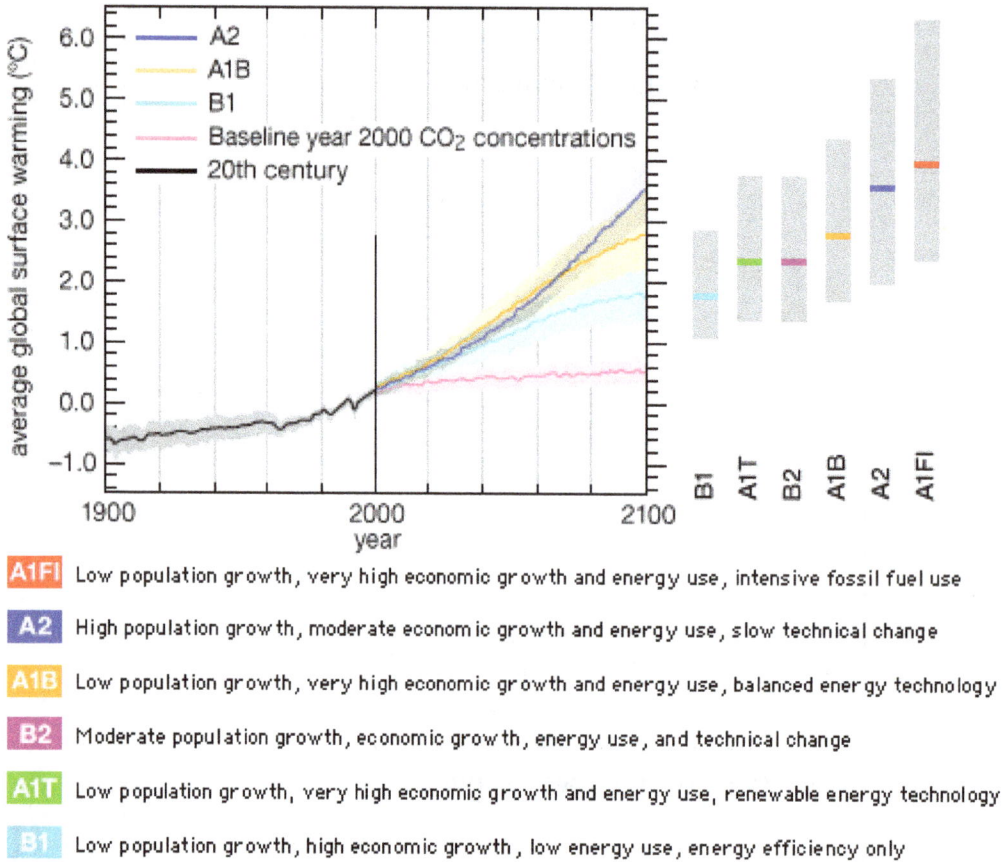

A2
A1B
B1
Baseline year 2000 CO$_2$ concentrations
20th century

average global surface warming ($^\circ$C)

B1   A1T   B2   A1B   A2   A1FI

year

A1FI   Low population growth, very high economic growth and energy use, intensive fossil fuel use

A2   High population growth, moderate economic growth and energy use, slow technical change

A1B   Low population growth, very high economic growth and energy use, balanced energy technology

B2   Moderate population growth, economic growth, energy use, and technical change

A1T   Low population growth, very high economic growth and energy use, renewable energy technology

B1   Low population growth, high economic growth, low energy use, energy efficiency only

Source: Climate Change 2007: The Physical Science Basis, Summary for Policymakers, Intergovernmental Panel on Climate Change

Figure 8. Global warming: estimations of past and future global warming. Source: Climate Change 2007: The Physical Science Basis, Summary for Policymakers, IPCC [8]

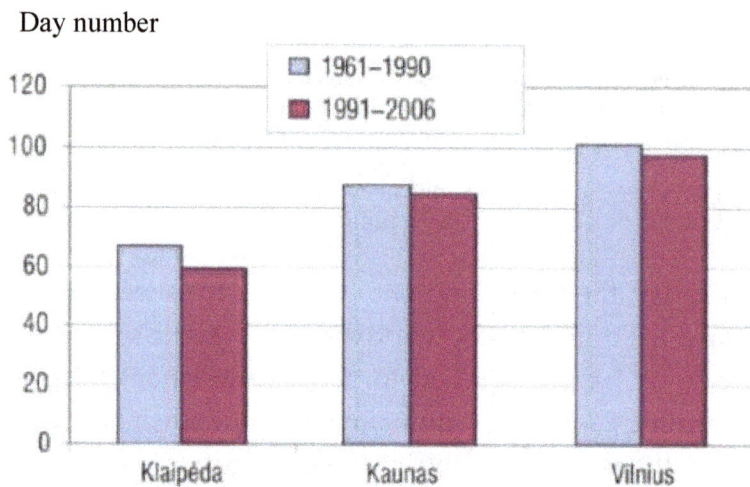

Day number

1961–1990
1991–2006

Klaipėda   Kaunas   Vilnius

Figure 9. Average number of days with snow cover in Klaipėda, Kaunas and Vilnius, Lithuania during 1961-1990 and 1991-2006. Source: Lithuanian Hydrometeorological Service under the Ministry of Environment [10]

Projected global average surface warming for 2020-2029 and the end of the 21st century (2090–2099) relatively to 1980–1999 are shown in figure 10. Projected warming in the 21st century shows scenario independent geographical patterns similar to those observed over the past several decades. Warming is expected to be the greatest over land and at the highest northern latitudes, and the least over the Southern Ocean and parts of the North Atlantic Ocean (Figure 10) [8].

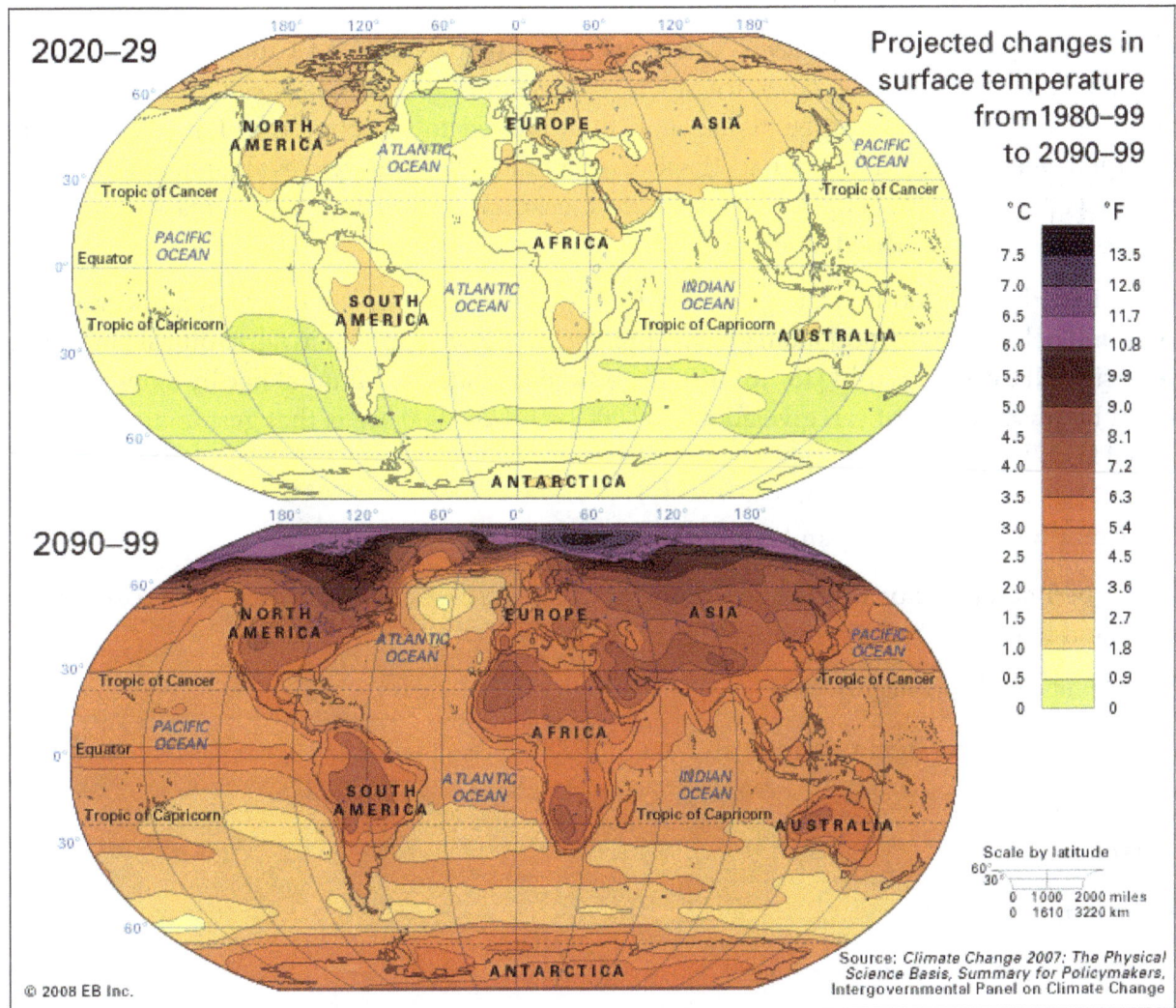

Figure 10. Projected changes in mean surface temperature by the late 21st century according to the A1B climate change scenario. All values for the period 2020-2029 and 2090–2099 are shown relatively to the mean temperature values for the period of 1980–1999. Source: Climate Change 2007: The Physical Science Basis, Summary for Policymakers, IPCC [8]

2. Material and methods

Lithuanian territory situated between 53°54′N and 56°27′N latitude, 20°56′E and 26°51′E longitude [53] occupies intermediate geographical position between west European oceanic climate and Eurasian continental climate. Cold air masses transfered from Arctic induce decrease of air temperatures which is reason of spring and early autumn frosts and of hard frost in winter time. Warm air masses from tropics are seldom which form thaws during winter and clear hot days during summer. Climate of the Lithuanian territory forms in different radiation and circulation conditions. Differences in these conditions hardly cross the boundaries of microclimatic differences; therefore, Lithuania belongs to western region of the Atlantic Ocean continental climatic area [45, 52] with average annual precipitation of 675 mm (572-907 mm) and temperature of 6-7°C [53, 54].

Phytotron vegetative pot experiments. The plastic pots (capacity of 5 L) substrata of turf (pH 6.0-6.5) were used. Till the emergence of white goosefoot *Chenopodium album* L. and one week after, pots were kept in a greenhouse, and then moved to the phytotron for 2 weeks. The emerged weeds were thinned out to 25 seedlings per pot. Results were evaluated after 21 days from weed emergence. Length of sprouts was measured (mm) and weed biomass (g per pot) was established oven-dried at 65°C.

Investigated weed genus white goosefoot *Chenopodium album* L. is widely spread in Europe, Asia and belongs to cosmopolitan group of plants. *C. album* is spread in agricultural lands and set–aside all over the world [15, 16]. White goosefoot *C. album* is annual hardly exterminated weed because one plant can give about 100,000 or till 200,000 seeds that germinate not all at once [15, 17].

2.1. Complex effect of CO_2 and temperature

The experimental factor was the environment of contrasting carbon dioxide (CO_2) concentrations and temperature level combinations.

Four levels of CO_2 concentration:

- 350 ppm (control treatment)
- 700 ppm
- 1500 ppm
- 3000 ppm

Two levels of temperature regimes:

- 21°C/17°C (control treatment)
- 25°C/21°C

CO_2 concentration, temperature regimes and their combinations were tested in the Phytotron vegetative pot experiments.

Concentration of CO_2 was regulated using CO_2 cylinder-reservoir controlled by CO_2 measurer "CO_2RT-5" (produced by Regin, Sweden). Photoperiod 16/8 h was achieved using high-pressure sodium (HPS) lamps SON-T Agro (Philips). The level of background radiation (PAR) made 170 micro-mol m^{-2} s^{-1}. PAR was measured with RF-100 Radiometer-Photometer with G.PAR-100 detector cell (produced by Sonopan, Poland).

2.2. Effect of UV-B radiation

Six levels of UV–B radiation (wavelength 290-320 nm) were tested:

- 0 kJ m^{-2} d^{-1}(control treatment)

- 1 kJ m^{-2} d^{-1}

- 3 kJ m^{-2} d^{-1}

- $^{-2}$ $^{-1}$

- 7 kJ m^{-2} d^{-1}

- 9 kJ m^{-2}d^{-1}

To generate the chosen UV–B radiation Medical lamps "Philips TL 40W/12 RS" UV–B were used.

2.3. Effect of ozone

Four levels of ozone concentrations were tested:

- 0 µg m^{-3} (control treatment)

- 120µg m^{-3}

- 240 µg m^{-3}

- 360 µg m^{-3}

The selected ozone concentration was reached using the ozone generator OSR-8 (Ozone Solutions, Inc.) 5 days per week, 7 hours per day. Ozone concentration was measured by the mobile ozone measuring equipment OMC-1108 (Ozone Solutions, Inc.).

2.4. Complex effect of UV-B radiation and ozone

Combination influence of two levels of ozone concentrations 120 µg m^{-3} and 360 µg m^{-3} with two levels of UV–B radiation: 3 kJ m^{-2}d^{-1}and 9 kJ m^{-2} d^{-1}was tested.

Experimental schema of UV–B radiation and ozone combinations is as follows:

- CT – control treatment, the plants were not exposed to either ozone or UV-B.

- O$_3$+O$_3$ – plants exposed to 120 µg m^{-3} ozone concentration and later to the supplemental 360 µg m^{-3} ozone concentration.

- O$_3$+UVB – plants exposed to 120 µg m^{-3} ozone concentration, and later to the supplemental 9 kJ m^{-2} d^{-1} UV-B radiation.

- UVB+UVB – plants exposed to 3 kJ m^{-2} d^{-1} UV-B radiation and later to the supplemental 9 kJ m^{-2} d^{-1} UV-B radiation.

- UVB+O$_3$ – plants exposed to 3 kJ m^{-2} d^{-1} UV-B radiation, and later to the supplemental 360 µg m^{-3} ozone concentration.

- $CT+O_3$ – plants exposed to 360 µg m^{-3} ozone concentration.

- CT+UVB-plants exposed to 9 kJ m^{-2} d^{-1} UV-B radiation.

2.5. Complex effect of ozone and temperature

Experimental photoperiod is 14/10 h. Ozone impact duration is 12 days.

Three levels of ozone concentrations:

- 20 µg m^{-3} (control treatment)

- 40 µg m

- 80 µg m^{-3}

Two levels of temperature regimes:

- 21°C/14°C (control treatment)

- 25°C/16°C

2.6. Complex effect of UV-B radiation and temperature

Experimental photoperiod is 14/10 h. Impact duration is 8 days.

Three levels of UV–B radiation:

- 0 kJ m^{-2} d^{-1} (control treatment)

- 2 kJ m^{-2} d^{-1}

- 4 kJ m^{-2} d^{-1}

Two levels of temperature regimes:

- 21°C/14°C (control treatment)

- 25°C/16°C

The experiments were conducted in three replications [18].

Data analysis. The collected data of the experiments were analysed by means of ANOVA. The treatment effects were tested for significance using the *Sigma Stat* software package [19] and the *Selekcija* software package [20].

3. Auto-ecological adaptability of weeds

Weeds are plants growing in undesirable places (i.e. crops and etc.) by human and competing with cultural plants for the growth factors and elements. Cultural plants can be counted as weeds if they are growing in crops of other cultural plants, for example, rye in wheat and etc. *Autoecology* is the branch of ecology which deals with individual species and their reactions to environmental factors. *Adaptivity* is the ability to react to change; adaptability allows the plant (weed) to function despite changes in the environment.

3.1. Complex effect of CO_2 and temperature

During the past decades the climate change and environment pollution became the important factors influencing the plant growth, development and productivity. The anthropogenic activity constantly changes the abiotical factors that surround us. The increasing air temperature, carbon dioxide, ozone, UV-B radiation and etc. are the factors constantly felt by the plants and their ability to adapt to the changing situation secures their productivity and agro-ecosystem stability [3].

Continued emissions of greenhouse gases will cause further warming and changes in all components of the climate system [7]. Limiting climate change will require substantial and sustained reductions of greenhouse gas emissions. Climate-carbon cycle coupling is expected to add carbon dioxide to the atmosphere as the climate system warms, but the magnitude of this feedback is uncertain. Based on current understanding of climate-carbon cycle feedback, model studies suggest that to stabilise at 450 ppm carbon dioxide could require that cumulative emissions over the 21st century would be reduced from the average of approximately 670 GtC (2460 GtCO$_2$) to approximately 490 GtC (1800 GtCO$_2$). Similarly, to stabilise at 1000 ppm, this feedback could require that cumulative emissions would be reduced from the model average of approximately 1415 GtC (5190 GtCO$_2$) to approximately 1100 GtC (4030 GtCO$_2$) [6, 8]. Depending on the scenario, about 15 to 40% of emitted CO_2 will remain in the atmosphere longer than 1,000 years [7]. This could result in the global climate change. Plants react to the increased concentration of CO_2, therefore this can trigger the processes of plant biomass accumulation [21].

Carbon dioxide as the carbon source used to synthesize the plant biomass is a very important abiotical factor in agriculture. Estimating the influence of CO_2 concentration for the growth of white goosefoot *Chenopodium album* L., the control treatment was compared with 350 ppm CO_2. Gradually CO_2 concentration was increased up to 700 ppm and 1500 ppm and it maximized the growth of white goosefoot and its biomass accumulation on the regular basis. When CO_2 concentration was increased up to 3000 ppm, the growth of *C. album* was reduced but still remained greater than CO_2 concentration in the control treatment (Figure 11). This means, that *C. album* can successfully adapt to the twice as great CO_2 concentrations, but it reaches the limit of the maximum growth. Other researchers [51] estimated also more intensive growth of other crop weed – *Parthenium hysterophorus* L. (whitetop weed) under a climate change scenario involving an elevated atmospheric CO_2 (550 μmol mol^{-1}) concentration. *P. hysterophorus* plants grew significantly taller (52%) and produced more biomass (55%) than under the ambient atmospheric CO_2 concentration (380 μmol mol^{-1}) [51].

The increasing concentration of atmospheric CO_2 is observed to increase plant photosynthesis and plant growth, which drives an increase of carbon storage in terrestrial ecosystems. However, plant growth is constrained by the availability of anthropogenic reactive nitrogen (Nr) in soils. This means that in some nitrogen-poor ecosystems, insufficient Nr availability will limit carbon sinks, while the deposition of Nr may instead alleviate this limitation and enable larger carbon sinks [22].

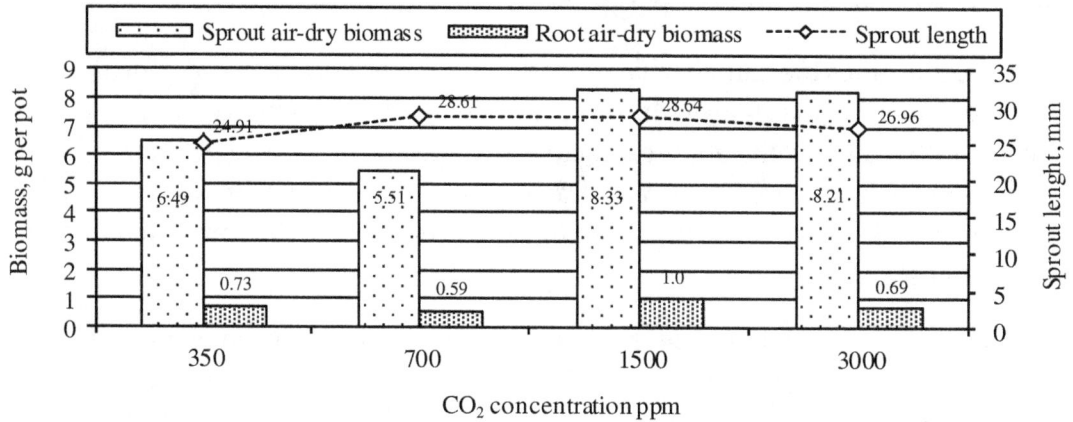

Figure 11. *Chenopodium album* L. white goosefoot accumulated biomass from a pot (g) and seedlings length (mm) under different CO_2 concentrations [23]

In addition to land use and climate-induced vegetation changes, CO_2 affects vegetation forcing indirectly, reducing transpiration from plants as stomata open less with increasing CO_2, resulting in localized atmospheric drying and warming [22, 24].

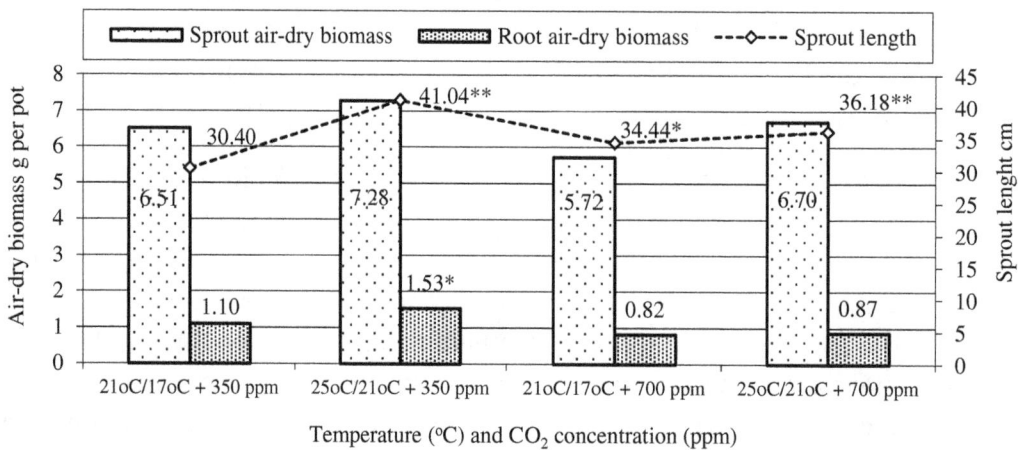

Note. * – significant differences in comparison with control treatment (21°C/17°C+350 ppm) at $P < 0.05$ and **-at $P < 0.01$.

Figure 12. *Chenopodium album* L. white goosefoot accumulated biomass from a pot (g) and seedlings length (mm) under different combinations of temperature regimes and CO_2 concentrations [23]

Continuing experiment under controlled phytotron conditions, complex impact of actual and forecasted CO_2 and temperature to the growth of white goosefoot *C. album* were tested (Figure 12). Increase of temperature to forecasted 25°C/21°C initiates growth of white goosefoot more quickly and accumulates its biomass more intensively. But the increase of CO_2 concentration up to 700 ppm at this forecasted temperature starts to inhibit the growth of white goosefoot and its roots (Figure 12). It was established that the most favourable conditions for *C. album* early growth were at higher temperature regime 25°C/21°C with both CO_2 concentrations. However, for the root growth initiation, especially, optimal conditions were with lower – 350 ppm – CO_2 concentration (Figure 12). Higher CO_2 concentration 700 ppm with actual lower temperature regime 21°C/17°C showed the negative influence on early growth of *C. album*.

3.2. Effect of UV-B radiation

The shorter are waves of the radiation, the greater is the effect of the ultraviolet radiation on living organisms [25, 26]. Molecule alternations and damages inevitably alter other processes: activity of genes, metabolism, intensity of photosynthesis which, consequently, influence the growth of the plant [27]. The reduction of the photosynthesis intensity due to the impact of UV-B radiation is related to slight conductance of stomata and the quantity of photosynthetic pigments [28]. Height and biomass of the majority of plant species have a tendency to reduce due to the UV-B radiation [28-30].

Experimental data showed that low UV-B radiation of 0, 1 and 3 kJ m^{-2} d^{-1} had a positive effect on early growth of *Chenopodium album* white goosefoot (Figure 13). *C. album* can accumulate up to 30% of biomass in excess in the background of 1 kJ m^{-2} d^{-1} UV-B radiation with reference to the control treatment. Increasing the intensity of UV-B radiation, the length of the *Chenopodium album* L. over-ground part was systemically decreasing. Consequently, as UV-B radiation constantly increased up to 9 kJ m^{-2} d^{-1}, the length of *C. album* over-ground part starting from 3 kJ m^{-2} d^{-1} gradually decreased by 28%. The over-ground part biomass of the *C. album* was decreasing respectively when increasing the UV-B radiation. The least over-ground part and root biomass of *C. album* were accumulated at the 9 kJ $m^{-2}d^{-1}$ UV-B radiation. Gradually increasing UV-B radiation till 9 kJ $m^{-2}d^{-1}$ *C. album* biomass decreased till 2 times compared it to the control treatment 0 kJ m^{-2} d^{-1} (Figure 13).

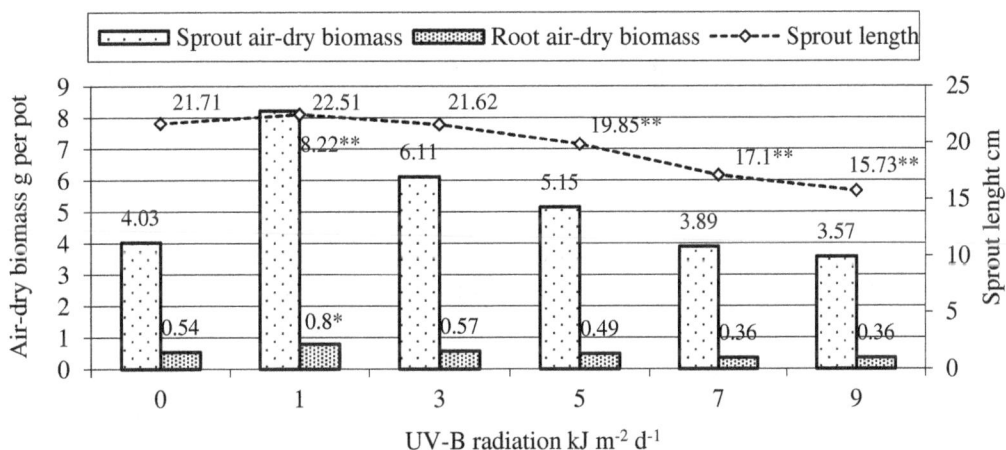

Note. Significant differences from control treatment (0 kJ m^{-2} d^{-1}) at * $P < 0.05$ and **-at $P < 0.01$

Figure 13. The above-ground and root biomass from a pot (g) and seedling length (mm) of *Chenopodium album* L. white goosefoot under different UV–B radiation (kJ m^{-2} d^{-1}) levels [31]

3.3. Effect of ozone

When the ozone layer in the stratosphere becomes thinner, the ozone concentration at the soil surface increases. Ozone concentration at the soil surface is also insecure to the plant development [3]. The ozone gas acts as strong oxidator in the plant cells and destabilizes the vital functions [32]. Short impact of ozone may cause various injuries to leaves, moreover, under the long-term continuous influence plants become less, the crop decreases, leaves are injured [33, 34]. Ozone adds to quicker senescence of plant leaves and their early fall. These processes are determined by the increase of free radicals in the plant cells [35, 36].

Evaluating the effect of ozone concentration on *C. album* white goosefoot growth, it has been established that the increasing ozone concentration had no statistically reliable impact on *C. album* growth, however, the tendency of over-ground part length (p=0.074) and air-dry biomass (p=0.958) decreasing was observed (Figure 14). The sprout height decreased by 15.4%, 16.8% and 2.0% in ozone concentration of 120 $\mu g\,m^{-3}$, 240 $\mu g\,m^{-3}$ and 360 $\mu g\,m^{-3}$ accordingly, compared it with the control treatment. *C. album* white goosefoot have lost 20.3%, 5.2% and 21.1% their sprout air-dry biomass at ozone concentration of 120 $\mu g\,m^{-3}$, 240 $\mu g\,m^{-3}$ and 360 $\mu g\,m^{-3}$ in respond to control treatment of 0 $\mu g\,m^{-3}$ of ozone, respectively.

Plants are known to suffer damage due to exposure to levels of ozone (O_3) above about 40 ppb [22, 37]. It is established that surface ozone detrimentally affects plant productivity [38]. Tropospheric ozone can also affect the natural uptake of CO_2 by decreasing plant productivity [22].

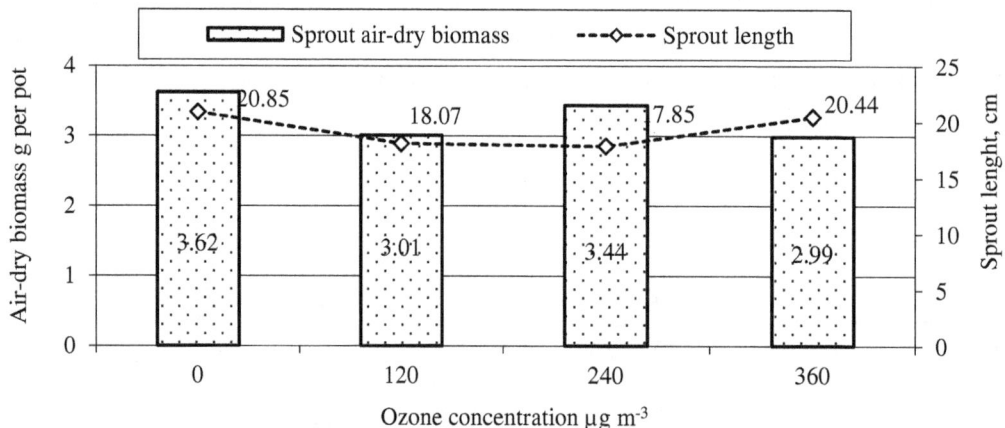

Figure 14. The *Chenopodium album* L. white goosefoot air-dry biomass and length of the over-ground part under the influence of ozone [31]

3.4. Complex effect of UV-B radiation and ozone

Ozone layer absorbs the greatest part of UV rays radiated by the Sun and other space bodies and protects plants and live organisms from their negative impact. Ozone depletion would increase the amount of ultraviolet light reaching the surface damaging terrestrial and marine ecosystems [22]. Since the beginning of the eight decade of the XX century the rapid breaking of the ozone layer in the stratosphere has been noticed as well as the increase of the intensity of UV radiation.

Ozone O_3 formed in the troposphere as a result of NO_x and volatile organic compound emissions reduces plant productivity, and therefore reduces CO_2 uptake from the atmosphere [22]. The depletion of the ozone layer is induced by the pollutants containing chlorine and bromine ions released into the environment [39]. The thickness of the ozone layer has the greatest impact on the flow of the UV-B radiation [25, 26].

During the complex research (Table 1) the negative impact of ozone and UV-B radiation on white goosefoot *C. album* growth increased in comparison to the impact of ozone (Figure 14) and UV-B (Figure 13) when effecting separately. *C. album* is unable to adapt to the increasing UV-B radiation and the intensifying complex impact of ozone and UV-B.

| Treatment | Over-ground plant part | | | | Root | |
| | After the first action | After the second action | | | | |
	Sprout length, mm	Sprout length, mm	Green biomass, g pot^{-1}	Air-dry biomass, g pot^{-1}	Green biomass, g pot^{-1}	Air-dry biomass, g pot^{-1}
#CT	44.17	48.38	54.04	12.37	10.19	1.77
O_3+O_3	50.33*	54.74 **	38.27*	11.50	10.60	1.90
O_3+UVB	51.00*	51.90*	46.69	10.82	12.73	1.98
UVB+UVB	47.53	47.98	45.83	12.10	9.60	1.75
$UVB+O_3$	51.13*	52.20*	35.58*	10.26	11.43	1.81
$CT+O_3$	49.63*	56.12**	37.84*	10.01	10.34	1.90
CT+UVB	51.83**	53.72*	50.11	12.92	14.86	2.07
P	0.017	0.001	0.024	0.398	0.071	0.820

Note. #CT – control treatment; significant differences from control treatment (#CT) at * $P < 0.05$ and **-at $P < 0.01$

Table 1. *Chenopodium album* L. white goosefoot biomass and length of the over-ground part after the exposure to both ozone and UV-B radiation [31]

3.5. Complex effect of ozone and temperature

Continuing research of ozone concentration, impact on white goosefoot *Chenopodium album* growth, complex effect of ozone concentration 20, 40 and 80µg m^{-3} and of actual and forecasted climate temperature regimes were evaluated (Figure 15). Increase of the ozone concentration from 20 µg m^{-3}, increased accumulation of *C. album* sprout air-dry biomass by 74% at 40 µg m^{-3} and by 68% at 80 µg m^{-3} and root air-dry biomass by 280% at 40 µg m^{-3} and by 200% at 80 µg m^{-3} in the actual climate temperature conditions 21°C/14°C. At forecasted climate higher temperature (25°C/16°C), the rising ozone concentration from 20 to 40 and 80µg m^{-3} increased the accumulation of *C. album* sprout air-dry biomass by 27-33% and root air-dry biomass by 23-50%, accordingly. Investigated ozone concentration and temperature regimes complex had no significant effect on *C. album* plant height (Figure 15). It was established that *C. album* is adapted to the actual and forecasted climate temperature and ozone concentration variations till 80 μg m^{-3} of ozone in the environment. Increasing ozone concentration further till 120, 240 and 360 μg m^{-3}, there was observed negative effect on *C. album* growth and abilities to adapt to higher ozone concentration were not determined (Figure 14).

The sprout root ratio of air-dry biomass changing concentration of ozone at different levels of temperatures showed, that *C. album* root growth was 3.5-5.8 times more intensive at forecasted than at actual climate temperature regimes (Lithuanian conditions). At actual climate temperature under ozone concentration of 20, 40 and 80 μg m^{-3} white goosefoot *C. album* sprout root ratio of air-dry biomass covered 39.8, 24.8 and 33.5 and at forecasted climate temperature covered 6.9, 7.1 and 6.1, accordingly.

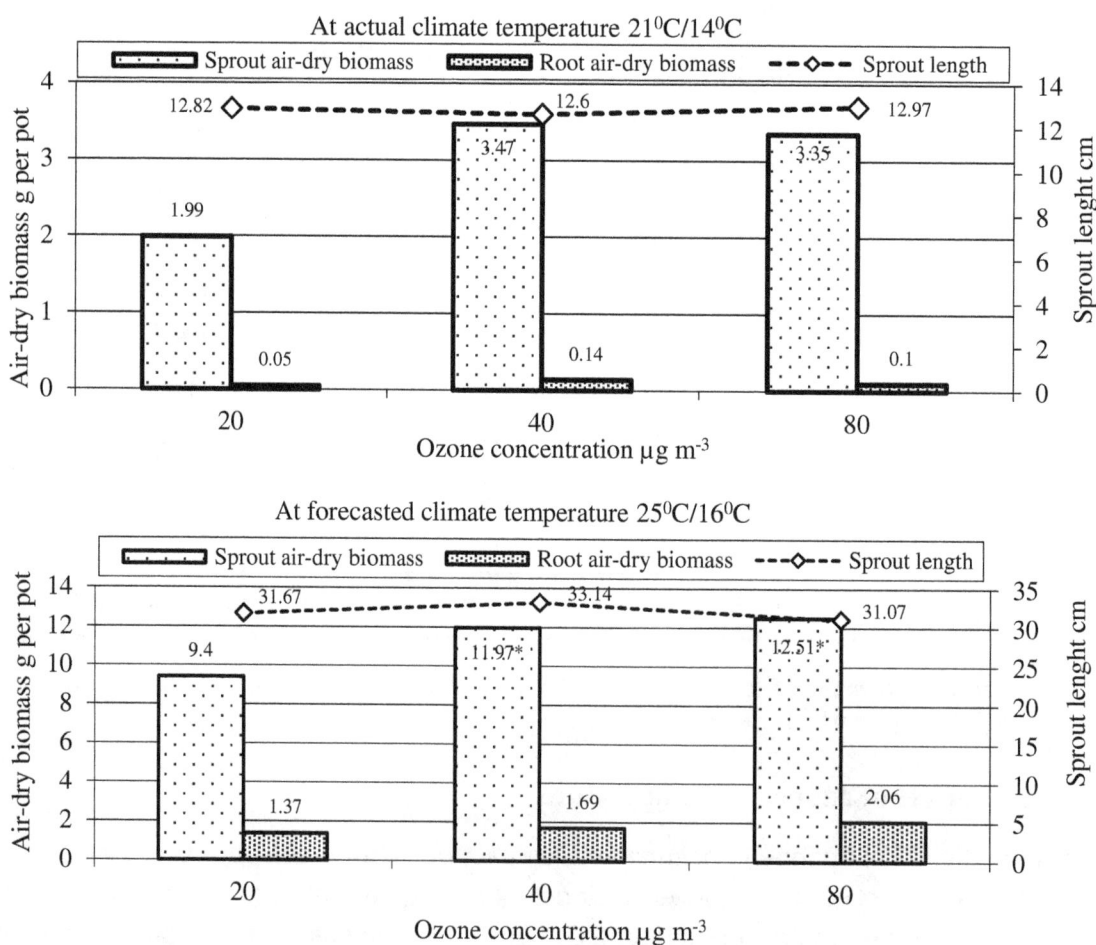

Note. * – significant differences from the control treatment (20 μg m^{-3}) at $P < 0.05$.

Figure 15. The influence of ozone on *Chenopodium album* L. white goosefoot growth at actual and forecasted climate temperature [40]

3.6. Complex effect of UV-B radiation and temperature

At changing climate conditions competitive abilities of plants are changing showing through new weed biological qualities. Increase of weed ability of over-wintering for weed species that during winter time traditionally were frosting at conventionally colder climate conditions [41, 42]. It was established that during winter time in winter wheat crop annual weeds, even some summer annual ones, had increased adaptivity of successful surviving winter frosts and accumulated higher one plant average mass by 5-6% during winter time; especially when the weather is favourable for prolonged development of weeds even at low density of perennial weeds in the crop [42-44]. Even short-time brief changes of meteorological conditions in crop during vegetation are inducing mechanism of plant/weed adaptivity. Namely, weed seed rain in the crop regularly intensified with increase of temperature and sunlight duration and vice versa [45, 46]. Under heavily polluted or dark cloudy skies, plant productivity may decline as the diffuse effect is insufficient to offset decreased surface irradiance [47]. Plants need a certain amount of UV-B radiation. They stimulate biochemical processes and inhibit to fast plant growing and slow accumulation of air-dry biomass [48]. Due to UV-B radiation height and air-dry biomass of many plant genus decrease [28, 29]. The intensity of UV-B radiation is determined by the seasson, day and night period and meteorological conditions. According to the data of Kaunas meteorological station and Palanga avia-meteorological station, the average UV-B radiation doses during clear summer days reach 2.1–2.5 kJ m^{-2} d^{-1} [49].

UV-B radiation	Over-ground part			Root		Over-ground part and root ratio	
	Height cm	Green mass g	Air-dry mass g	Green mass g	Air-dry mass g	Green mass	Air-dry mass
Actual climate temperature 21°C/14°C							
0 kJ m^{-2} d^{-1}	7.10	28.27	1.89	2.35	0.25	12.0	7.6
2 kJ m^{-2} d^{-1}	5.53**	11.63**	1.01*	1.12*	0.13*	10.4	7.8
4 kJ m^{-2} d^{-1}	6.31**	5.02**	0.53**	0.88*	0.09*	5.7	5.9
P	0.001	0.001	0.002	0.040	0.013	–	–
Forecasted climate temperature 25°C/16°C							
0 kJ m^{-2} d^{-1}	21.36	68.89	6.05	2.92	0.54	23.6	11.2
2 kJ m^{-2} d^{-1}	14.47**	34.73**	3.22**	1.89	0.28**	18.4	11.5
4 kJ m^{-2} d^{-1}	12.22**	15.38**	1.63**	0.94*	0.16**	16.4	10.2
P	0.001	0.001	0.001	0.017	0.002	–	–

Note. * – significant differences from the control treatment (0 kJ m^{-2} d^{-1}) at $P < 0.05$ and ** – at $P < 0.01$.

Table 2. The influence of UV-B radiation on white goosefoot C. album growth at actual and forecasted climate temperature [40]

The next experiment increasing UV-B radiation intensity till 4 kJ m^{-2} d^{-1} showed significant negative influence on white goosefoot *C. album* development already at UV-B radiation 2 kJ m^{-2} d^{-1} at both-actual 21°C/14°C and forecasting 25°C/16°C climate temperature regimes (Table 2).The over-ground green biomass of *C. album* effected by UV-B radiation of 4 kJ m^{-2} d^{-1}decreased 5.6 and 4.5 times at actual and forecasted climate temperature respectively compared it with the control treatment (UV-B radiation 0 kJ m^{-2} d^{-1}), while root green biomass decreased 2.7 and 3.1 times, accordingly. *C. album* over-ground part and root air-dry biomass accumulation decreased nearly twice (1.9 times) already at UV-B radiation 2 kJ m^{-2} d^{-1} at both temperature regimes and declined till 3.6-3.7 and 2.8-3.4 times at UV-B radiation 4 kJ m^{-2} d^{-1}, accordingly. Increasing intensity of UV-B radiation, *C. album* height growth inhibited significantly as well. The highest evaluated UV-B radiation in the experiment decreased plant height by 11% at actual climate temperature and by 43% at forecasted warmer climate temperature. Received data of experiment confirmed that *C. album* is sensitive to UV-B radiation in actual colder and forecasted warmer temperature regimes. *C. album* plant over-ground part and root green and air-dry biomass ratio with increase of UV-B radiation regularly decreased. It could be result of plant protection mechanism activation intensifying transpiration process.

4. Conclusions

1. Plant ability to survive under unfavourable conditions depends upon the intensity and character of the unfavourable factors. Abiotical factors of low intensity influencing plants induce weed growth, however, weed growth is regularly smothered as their intensity increases.

2. Increase of CO_2 concentration positively affected the early growth of white goosefoot *Chenopodium album* L. and reached the optimum at 1500 ppm. Higher temperature regime 25°C/21°C compared with 21°C/17°C compounded more favourable conditions for *C. album* early growth at both – 350 ppm and 700 ppm – CO_2 concentrations. White goosefoot successfully adapts even to several times increased concentration of CO_2.

3. Minor UV-B radiation concentrations 1-3 kJ m^{-2} d^{-1} induced *C. album* growth; however, the increasing UV-B radiation (5-9 kJ m^{-2} d^{-1}) reliably decreased both the length of the over-ground part and the biomass of white goosefoot.

4. Increasing ozone concentration to 120, 240 and 360 μg m^{-3} had a tendency to suppress *C. album* growth by 2-14% of its sprout length and by 5-17% of its accumulated air-dry biomass. However, complex investigation of ozone and temperature showed that *C. album* is adapted to actual and forecasted climate temperature and ozone concentration variations till 80 μg m^{-3} of the environment.

5. Complex investigation of UV-B radiation and temperature showed significantly negative influence on *C. album* growth and biomass accumulation already at UV-B radiation 2 kJ m^{-2} d^{-1} of both actual 21°C/14°C and forecast 25°C/16°C climate temperature regimes.

6. Joint action of ozone and UV-B radiation on *C. album* growth increased negative effect relatively to separate impact of ozone and UV-B. White goosefoot in early growth stage is unable to adapt to increasing UV-B radiation (>3 kJ m^{-2} d^{-1}) and the intensifying complex impact of ozone and UV-B.

7. The experimental results suggest that in long-term (more than 30 years) time period weeds are well adapted to changing climate conditions and will become more competitive in temperate climate zone. For successful weed control in the crop of agricultural plants present weed control methods and strategy should be reviewed and improved adapting them to threats of global warming and climate change.

Acknowledgements

The Lithuanian State Science and Studies Foundation as a part of the research project "Complex effect of anthropogenic climate and environment changes on the forest and agro ecosystem flora" supported this research.

We would like to thank Vilma Pilipavičienė for the manuscript English reviewing linguistically.

Author details

Vytautas Pilipavičius*

Address all correspondence to: vytautas.pilipavicius@asu.lt

Aleksandras Stulginskis University, Faculty of Agronomy, Institute of Agroecosystems and Soil Sciences, Akademija, Lithuania

References

[1] Biota ir aplinkos kaita. Žalakevičius M. (ed.). Vilniaus universitetas Ekologijos institutas. Vilnius; Vol.1. 2007. p.239.

[2] Vidickienė D., Melnikienė R., Gedminaitė-Raudonė Ž. Climate Change: Influence on Agriculture and Forestry in Lithuania. Globalization, the European Union's development, regionalization processes: Changes and new challenges 2011; 4(32) 82-89.

[3] Pilipavičius V. Weed spreading regularity and adaptivity to abiotical factors: summary of the review of scientific works presented for dr. habil. procedure. Lithuanian University of Agriculture, Kaunas; 2007. p.30.

[4] Pilipavičius V., Romaneckienė R., Romaneckas K. Crop stand density enhances competitive ability of spring barley (*Hordeum vulgare* L.). Acta Agriculturae Scandinavica. Section B, Soil and plant science 2011; 61(7) 648-660.

[5] Pilipavičius V., Romaneckienė R., Ramaškevičienė A., Sliesaravičius A., Burbulis N., Duchovskis P. 2005. *Chenopodium album* and *Rumex crispus* seedling biomass, root and sprout formation dependence on different abiotic factors and their combinations. Agronomijas vēstis 2005; 8 156-161.

[6] Climate Change 2007: The Physical Science Basis. Contribution of Working Group I to the Fourth Assessment Report of the Intergovernmental Panel on Climate Change [Solomon S., Qin D., Manning M., Chen Z., Marquis M., Averyt K.B., Tignor M., Miller H.L. (eds.)]. Cambridge University Press, Cambridge, United Kingdom and New York, NY, USA, 2007. 996 pp.

[7] IPCC, 2013: Summary for Policymakers. The Physical Science Basis. Contribution of Working Group I to the Fifth Assessment Report of the Intergovernmental Panel on Climate Change [Stocker T.F., Qin D., Plattner G.-K., Tignor M., Allen S.K., Boschung J., Nauels A., Xia Y., Bex V., Midgley P.M. (eds.)]. Cambridge University Press, Cambridge, United Kingdom and New York, NY, USA. 2013.

[8] IPCC, 2007: Summary for Policymakers. In: Climate Change 2007: The Physical Science Basis. Contribution of Working Group I to the Fourth Assessment Report of the Intergovernmental Panel on Climate Change [Solomon S., Qin D., Manning M., Chen Z., Marquis M., Averyt K.B., Tignor M., Miller H.L. (eds.)]. Cambridge University Press, Cambridge, United Kingdom and New York, NY, USA. 2007.

[9] Climate Change 2007: Synthesis Report. An Assessment of the Intergovernmental Panel on Climate Change. This summary, approved in detail at IPCC Plenary XXVII (Valencia, Spain, 12-17 November 2007), represents the formally agreed statement of the IPCC concerning key findings and uncertainties contained in the Working Group contributions to the Fourth Assessment Report. Based on a draft prepared by: Bernstein L. et al. 2007.

[10] Lietuvos gamtinė aplinka, būklė, procesai ir raida 2008. Bukantis A., Gedžiūnas P., Giedraitienė J., Ignatavičius G., Jonynas J., Kavaliauskas P., Lazauskienė J., Reipšleger R., Sakalauskienė G., Sinkevičius S., Šulijienė G., Žilinskas G., Valiukevičius G. Aplinkos apsaugos agentūra, Vilnius, 2008. 238.

[11] Jones P.D., Moberg A. Hemispheric and large scale surface air temperature variations: An extensive revision and an update to 2001. Journal of Climate 2003; 16 206-223.

[12] Bockris J. Global Warming. In: Global Warming, edited by Harris S.A. Rijeka: InTech; 2010. p.159-220. Available from: http://www.intechopen.com/books/global-warming/global-warming-(accessed 7 May 2014)

[13] Klimatas. Lietuva: kompiuterinė enciklopedija. Paltanavičius S., Gudžinskas Z. Available from: http://mkp.emokykla.lt/enciklopedija/lt/straipsniai/zeme/klimatas/klimatoprognozes (accessed 6 May 2014)

[14] Rimkus E., Bukantis A. Climate change in Lithuania. International scientific confer-
      ence: Climate change and forest ecosystems. Vilnius, 22-23 October, 2008. 141-142.

[15] Aleksandravičiūtė B., Apalia D., Brundza K. et al. (1961) Lietuvos TSR flora / Flora of
      Lithuania. Vol.3. Minkevičius A. (ed.). Mokslas, Vilnius, Lithuania. 1961.

[16] Holm L.G., Pancho J.V., Herberger J.P., Plucknett D.L. (1979) Geographical Atlas of
      World Weeds. John Wiley & Sons, New York, USA. 1979.

[17] Grigas A. Lietuvos augalų vaisiai ir sėklos. Mokslas, Vilnius, Lithuania. 1986.

[18] Pilipavičius V., Lazauskas P. Optimal number of observation, treatment and replica-
      tion in field experiments. African Journal of Agricultural Research 2012; 7(31)
      4368-4377.   Available   from: http://www.academicjournals.org/article/arti-
      cle1380817068_Pilipavicius%20and%20Lazauskas.pdf (accessed 30 April 2014)

[19] SPSS Science. SigmaStat® Statistical Software Version 2.0. User's Manual. USA. 1997.

[20] Tarakanovas P. A new version of the computer programme for trial data processing
      by the method of analysis of variance. Žemdirbystė-Agriculture 1997; 60 197-213.

[21] Rogers H.H., Runion G.B., Krupka S.V. Plant responses to atmospheric CO_2 enrich-
      ment with emphasis on roots and rhizosphere. Environmental Pollution 1994; 83
      155-189.

[22] IPCC, 2013: Climate Change 2013: The Physical Science Basis. Contribution of Work-
      ing Group I to the Fifth Assessment Report of the Intergovernmental Panel on Cli-
      mate Change [Stocker T.F., Qin D., Plattner G.-K., Tignor M., Allen S.K., Boschung J.,
      Nauels A., Xia Y., Bex V., Midgley P.M. (eds.)]. Cambridge University Press, Cam-
      bridge, United Kingdom and New York, NY, USA, 1535 pp.

[23] Pilipavičius V., Romaneckienė R., Ramaškevičienė A., Sliesaravičius A. The effect of
      CO_2 and temperature combinations on Chenopodium album L. early growth. Agrono-
      my Research 2006; 4(Special issue) 311-316.

[24] Joshi M., Gregory J. Dependence of the land-sea contrast in surface climate response
      on the nature of the forcing. Geophysical Research Letters 2008; 35(24). L24802, doi:
      10.1029/2008GL036234.   Available   from: http://onlinelibrary.wiley.com/doi/
      10.1029/2008GL036234/pdf (accessed 2 May 2014)

[25] Helsper J.P.F.G., Ric de Vos C.H., Mass F.M., Jonker H.H., van der Broeck H.C., Jordi
      W., Pot C.S., Kleizer L.C.P., Schapendonk A.H.C.M. Response of selected antioxi-
      dants and pigments in tissues of Rosa hybrida and Fuchsia hybrida to supplemental
      UV-A exposure. Physiologia Plantarum 2003; 117 171-178.

[26] Krizek T.D. Influence of PAR and UV-A in determining plant sensitivity and photo-
      morphogenic responses to UV–B radiation. Photochemistry and Photobiology 2004;
      79 307-315.

[27] Brosche M., Strid A. Molecular events following perception of ultraviolet-B radiation by plants. Physiologia Plantarum 2003; 117 1-10.

[28] Ambasht N.K., Agrawal M. Influence of supplemental UV-B radiation on photosynthetic characteristics on rice plants. Photosynthetic 1997; 34 401-408.

[29] Correia C.M., Torres Pereira M.S., Torres Pereira J.M. Growth, photosynthesis and UV-B absorbing compounds of Portuguese barbela wheat exposed to UV-B radiation. Environmental Pollution 1999; 104 383-388.

[30] Mazza C.A., Battista D., Zima A.M., Szwarcberg-Bracchitta M., Giordano C.V., Acevedo A., Scopel A.L., Ballare C.L. The effects of solar ultraviolet-B radiation on the growth and yield of barley are accompanied by increased DNA damage and antioxidant responses. Plant, Cell and Environment 1999; 22 61-70.

[31] Pilipavičius V., Romaneckienė R., Ramaškevičienė A., Sliesaravičius A. Effect of UV-B radiation, ozone concentration and their combinations on *Chenopodium album* L. early growth adaptivity. Žemdirbystė-Agriculture 2006; 93(3) 99-107.

[32] Saitanis C.J., Riga-Karandinos A.N., Karandinos M.G. Effects of ozone on chlorophyll and quantum yield of tobacco (*Nicotiana tabacum* L.). Chemosphere 2001; 42(8) 945-953.

[33] Krupa S.V., Grunghoge L., Jager H.J., Nosal M., Legge A.H., Hanawald K. Ambient ozone and adverse crop response: a unified view of cause and effect. Environmental Pollution 1995; 99 398-405.

[34] Reddy K.R., Hodges H.F. Climate change and global Crop Productivity. CABI Publishing. 2000. 472 p.

[35] Farage P.K., Long S.P., Lechner E.G., Baker N.R. The sequence of changes within the photosynthetic aparatus of wheat following shortterm exposure to ozone. Plant Physiology 1991; 95 529-535.

[36] Fumagalli I., Gimeno B.S., Velissariou D., de Temmermand L., Millse G. Evidence of ozone-induced adverse effects on crops in the Mediterranean region. Atmospheric Environment 2001; 35(14) 2583-2587.

[37] Ashmore M.R. Assessing the future global impacts of ozone on vegetation. Plant Cell and Environment 2005; 28 949-964.

[38] Fishman J., Creilson J.K., Parker P.A., Ainsworth E.A., Vining G.G., Szarka J., Booker F.L., Xu X. An investigation of widespread ozone damage to the soybean crop in the upper Midwest determined from ground-based and satellite measurements. Atmospheric Environment 2010; 44 2248-2256.

[39] Rozema J., van de Staaij J., Bjorn L. O., Caldwell M. UV–B as an environmental factor in plant life: stress and regulation. Trends in Ecology & Evolution 1997; 12 22-28.

[40] Romaneckienė R., Pilipavičius V., Romaneckas K. The influence of ozone and UV-B radiation on fat-hen (*Chenopodium album* L.) growth in different temperature conditions. Žemdirbystė-Agriculture 2008; 95(4) 122-132.

[41] Pilipavičius V., Romaneckas K., Gudauskienė A. Weed Seedling Over-Wintering and Vegetation Dynamics in Organically Grown Winter Wheat Spelt Crop under Climate Changing Conditions. Rural development 2013: the 6th international scientific conference, 28-29 November, 2013, Aleksandras Stulginskis university, Akademija, Kaunas district, Lithuania: proceedings. Vol. 6, book 2 (2013), p. 208-212.

[42] Pilipavicoius V. Herbicides in winter wheat of early growth stages enhance crop productivity. In: Herbicides-Properties, Synthesis and Control of Weeds. Editor Hasaneen M.N. Rijeka: InTech. 2012. p. 471-492. Available from: http://cdn.intechopen.com/pdfs-wm/25635.pdf (accessed 9 May 2014)

[43] Pilipavičius V., Aliukonienė I., Romaneckas K. Chemical weed control in winter wheat (*Triticum aestivum* L.) crop of early stages of development: I. Crop weediness. Journal of Food, Agriculture & Environment 2010; 8(1) 206-209.

[44] Pilipavičius V., Aliukonienė I., Romaneckas K., Šarauskis E. Chemical weed control in the winter wheat (*Triticum aestivum* L.) crop of early stages of development: II. Crop productivity. Journal of Food, Agriculture & Environment 2010; 8(2) 456-459.

[45] Pilipavičius V. Weed seed rain dynamics and ecological control ability in agrophytocenosis. In: Herbicides-Advances in Research. Edited by Price A. J. and Kelton J. A. Rijeka: InTech. 2013. p. 51-83. Available from http://cdn.intechopen.com/pdfs-wm/43456.pdf (accessed 12 May 2014)

[46] Pilipavičius V. Piktžolių sėklų byrėjimo priklausomumas nuo meteorologinių faktorių / Dependence of Weed Seed Falling on Meteorological Factors, Precipitation and Sunlight Duration. Vagos 2002; 53(6) 17-21.

[47] UNEP, 2011: Integrated assessment of black carbon and tropospheric ozone: Summary for decision makers. United Nations Environment Programme and World Meteorological Association. 2011. 38 p.

[48] Wci G., Zheng Y., Slusser J.R., Heisler G.M. Impact of enhanced ultraviolet – B radiation on growth and leaf photosynthetic reaction of soybean (*Glicine max*). Physiologia Plantarum 2003; 52 353-362.

[49] Jonavičienė R. Ultravioletinės saulės spinduliuotės matavimai Lietuvos hidrometeorologijos tarnyboje. Meteorologija ir hidrologija Lietuvoje: raida ir perspektyvos. Vilnius, 2005. 48-49.

[50] Verlinden M., de Boeck H.J., Nijs I. Climate warming alters competition between two highly invasive alien plant species and dominant native competitors. Weed Research 2014; 54(3) 234-244.

[51] Shabbir A., Dhileepan K., Khan N., Adkins S.W. Weed-pathogen interactions and elevated CO_2: growth changes in favour of the biological control agent. Weed Research 2014; 54(3) 217-222.

[52] Basalykas A., Bieliukas K., Chomskis V. Lietuvos TSR fizinė geografija / Physical geography of Lithuania. Vilnius: Mokslas; 1958.

[53] Visuotinė Lietuvių enciklopedija / Universal Lithuanian Encyclopedia. Lietuva / Lithuania. Klimatas / Climate. Vaitekūnas S. *et al.* (ed.). Vilnius: Mokslo ir enciklopedijų leidybos institutas; 2007. Vol.12. 47-57.

[54] Pilipavičius V., Grigaliūnas A. Lithuanian Organic Agriculture in the Context of European Union. In: Organic Agriculture towards Sustainability. Edited by Pilipavičius V. Rijeka: InTech. 2014. p. 89-121. DOI: 10.5772/58352. Available from: http://cdn.intechopen.com/pdfs-wm/46459.pdf (accessed 30 May 2014)

[55] Singh B.R., Singh O. Study of Impacts of Global Warming on Climate Change: Rise in Sea Level and Disaster Frequency. In: Global Warming-Impacts and Future Perspective. Edited by Singh B.R. Rijeka: InTech. 2012. p. 93-118. DOI: 10.5772/50464. Available from: http://www.intechopen.com/books/global-warming-impacts-and-future-perspective/study-of-impacts-of-global-warming-on-climate-change-rise-in-sea-level-and-disaster-frequency (accessed 19 June 2014)

[56] Lazauskas P., Pilipavičius V. Agroekologija / Agroecology. Lietuvos žemės ūkio universitetas. Akademija [i.e.Klaipėda]: IDP Solutions. 2008. 140 p.

[57] Heinrich D., Hergt M. Ekologijos atlasas. Vilnius: Alma littera. 2000. 279 p.

2

A Study on Economic Impact on the European Sardine Fishery due to Continued Global Warming

M. Dolores Garza-Gil, Manuel Varela-Lafuente,
Gonzalo Caballero-Míguez and Julia Torralba-Cano

1. Introduction

The global warming is one of the biggest issues now facing humanity, including its repercussions on climate change. The climate change is directly or indirectly attributable to human activity and alters the composition of the atmosphere, along with the natural variability observed during comparable periods of time [1]. On a global level, human activity has generated a loss in biodiversity due, among other factors, to changes in the way we use the land, the pollution of soil and water, the diversion of water to intensively farmed ecosystems (intensive agricultural and livestock farming) and urban systems, the introduction of non-autochthonous species and the exhaustion of the ozone layer. In particular, the rate of biodiversity loss is currently greater than that of natural extinction [2]. In this context, climate change is contributing towards increasing this loss of biological biodiversity [3]. It is considered that the ongoing increase in the levels of greenhouse gases is one of the main causes, along with natural factors, of climate change [2]. The concentrations of greenhouse gases have been increased since the industrial age due fundamentally to the use of fossil fuels and to changes in soil use and cover; generating, among other effects, increases in land and sea surface temperatures, changes in precipitation (both spatial and seasonal), increases in sea levels and changes in the frequency and intensity of some climate phenomena (hurricanes, intensive precipitations, heat weaves). [4-5]

The scientific certainty that climate change is a reality [6] and, therefore, that it is necessary to live alongside it, leads us to take action aimed at reducing greenhouse gasses compatible with other action which aims to study the possibilities of adapting to these new environmental conditions that are going to characterise the planet in the coming decades. The new scenarios generated by this global phenomenon generate negative impacts on the majority of Spain's ecosystems, and also on productive activities; but opportunities may also arise (for example, for tourism in northern Spain; [7]) and could be tapped into, if a certain level of scientific knowledge and prospective capacity with regard to the evolution of environmental conditions in the regions and their effects on natural systems over the next decades were reached.

In one of its latest reports, the Inter-governmental Panel on Climate Change (IPCC) estimates that, on a global level, the average temperature of the land surface could increase by between 1.4 and 5.8°C by the end of the 21st century (a greater increase in higher latitudes that in the tropics), that there will be an increase in the sea surface temperature, that terrestrial regions will undergo a higher increase than the oceans, that the sea will rise by between 0.09 and 0.88m, and that there will be an increase in the occurrence of extreme phenomena (heavy rainfall, hurricanes, etc.), although with regional differences (IPCC, 2013). Proximity to one or other extreme of the intervals will depend fundamentally on society's capacity to mitigate the effects of climate change.

The different scientific reports published by national [8-9] and international [6] bodies on global warming leave no doubt as to Spain's vulnerability to its effects. However, this knowledge is not sufficient to describe the behaviour of the various microclimates in different areas of Spain and their interaction with different ecosystems throughout the country. Northern Spain, to where the fishery that we are going to look at in this paper belongs, is located in medium latitudes, framed within an area of atmospheric circulation where western and northern winds prevail, and is the first point Atlantic perturbations reach with respect to the rest of Spain. However, at the same time, this zone is influenced by different air masses with very different thermodynamic characteristics. Specifically, masses of warm, humid air arrive, like tropical maritime masses, as well as masses of air which, as they come from higher latitudes, have the common characteristic of being cold, although with a different humidity content [10].

In the specific case of the oceans, one of their significant characteristics is that they store a much greater quantity of energy than the atmosphere, and thus the possible effects of global warming would have a greater impact on marine ecosystems than on land ecosystems [11-15]. Based on the last reports from the IPCC, the global warming will generate changes in the intensity and structure of oceanic currents, produce alterations in marine organisms and shoreline, among other effects. Such shocks on the oceans probably will have important repercussions on the fisheries [16-20]. The projected impacts of global warming on fisheries and aquaculture are negative on a global scale. These impacts of climate change and ocean acidification are generally exacerbated by other factors such as overfishing, habitat loss and pollution. This is contributing to an increase in the number of dead zones in the oceans, as well as to an increase in harmful algal blooms [6]. The fishing activity is conditioned by the natural characteristics of the marine environment. Therefore, this economic activity is located between the most affected by global warming, along with tourism and agriculture ([7], [21]-[25]). It is highly probable that global warming will generate impacts on the intensity and disposition of oceanic currents, and the effect of this, among others, will generate increasing in surface sea temperature, variations in acidification and salinity levels and variationson ([26], [6]-[7], [26]). The impacts will differ according to ecosystems and coastal or ocean zones, and will affect different groups of organisms, from phytoplankton and zooplankton to fish and algae [9]. Among these organisms, small pelagic marine species will be among those most affected by the effects form the warmin gf seas and oceans, due to their high level of instability and sensitivity to environmental impacts [27-29]. Therefore, any sea surface temperature variation will have repercussions to a greater or lesser extent on the ecosystems and these species' reproduction levels [17], especially in warming scenarios, and so any significant modification in the biomass levels can affect fishermen's net profits.

In the northern Spain zone, an increase in the sea surface temperature of 0.2°C per decade over the last forty years has been observed [10]. An increase has also been observed in the sea level of between 2 and 2.5cm per decade, similar data to the global increase in the Atlantic Ocean, although it has accelerated in the last twenty years [10]. Furthermore, a decrease in the intensity and duration of upwelling due to changes in wind patterns has occurred (the duration of the period favourable for upwelling has decreased significantly, by 30%, and their intensity by 45% in the last forty years; [10]). For the northern zone overall, the IPCC foresees an rise in the sea level of between 0.5 and 1.4 metres, an increase of the sea surface temperature of between 1 and 3°C and a drop in the pH of around 0.35 units [6].

The small pelagic species are target species for the majority of the Spanish fleet, in general, and for the north-west in particular. The study case chosen is the sardine fishery in zones VIIIc and IXa delimited by the International Council for the Exploration of the Sea (ICES), due to it being a hugely important pelagic species for the European fisheries sector. The potential economic impacts of global warming on this fishery will be analysed. For it, the oceantemperature is introduced into the bio-economic management problem. The sea surface allows gathering evidence of the global warming and its effects on marine ecosystems, which are the bases of fish natural growth functions. Other variables, such as the frequency and intensity of rainfall, acidity, dissolved carbon and salinity, are also prone to experience environmental changes by the climate change. However there is a high level of correlation between all of these variables: greater frequency and intensity of rainfall, acidity, salinity and dissolved carbon when the sea surface is increasing.

In this chapter, the evolution of sardine biomass is showed, based on the changes of the stock's growth function with respect to variations in marine ecosystem characteristics through the sea surface temperature and using econometric techniques and the Theory of the Optimal Control. Correlations between the sea surface temperature and the sardine biomass are tested. The econometric techniques will allow define the significant parameters and variables that could be needed to include in the management problem. The Optimal Control Theory will allow solve the bio-economic management problem and obtain the values for the sardine biomass and catches according to the forecast scenarios for the surface sea temperature by the IPCC. In addition, we will analyse impacts on the economic yield of the fishery deriving from a possible change in the temperature conditions of the ecosystem. A strong statistical relation is showed between the sea surface temperature and the sardine biomass. In addition, the results show thatif the sea surface temperature trend in these fishing grounds continues to show warming (as it is expected), the European sardine biomass, the catches and the expected profits will drop over the medium- and long-term.

2. Material and method

The Ibero-atlantic sardine (*Sardina pilchardus*) fishery is a stock resource shared by Spain and Portugal, in the waters that surround the Iberian Peninsula between the Bay of Biscay and the Strait of Gibraltar. There are historical references to this fishery dating back to the 13th and 14th centuries, made by both Spain [30] and Portugal [31]. From its beginnings, this fishery has been significantly relevant for the fishery sectors of both countries as well as for the processing industry (salting and canning industries).

This stock is exploited majority by the purse seine gear (99% of landings [32]). The Spanish fleet using this gear in the Atlantic is composed by 491 vessels, of which 346 are involved in Cantabrian waters and the remaining 146 are fishing in the grounds of the Canary Islands and the Bay of Cadiz [33]. It is one of the more numerous fleets of the Spanish total, behind only of the artisanal fleet, and it targets small pelagic species, among which are the sardine, horse mackerel, mackerel and the anchovy ([7],[33]-[34]). The vessels involved in this fisheryare relatively homogenous vessels insofar as their technical characteristics are concerned, with an average fishing capacity of 34.2 GT, 151.8 Kw per vessel and 21m of length size. The average life of the fleet is 20 years, with a crew of 8 per vessel. The Spanish purse seine fleet in the North Atlantic grounds use nets made from synthetic materials, hydraulic haulers and electronic fish detectors. The Spanish fleet that operates in the purse seine fishery licensed to catch sardine in this fishery is 241 vessels in 2012 [35].

2.1. The sea surface temperature

The Ibero-atlantic sardine recruitment processes could be governed by localised oceanographic conditions, but also by climatic events of a global nature. The growing influence of subtropical climatic components, such as El Niño, would increase the sea surface temperature, while the frequency and intensity of the average coastal upwellings on the west coast of the Iberian Peninsula are reduced, probably meaning that such a combination would cause the fishery's productivity to decrease [36-37]. Therefore, in this paper, we are going to establish some of the relationships between environmental alterations that could be caused by global warming, through the sea surface temperature, and the sardine fishery. We will analyse a bio-economic model which will include alterations in the sea surface temperature, and we will estimate their effects on fish population dynamics.

With respect to the sea surface temperature (SST), the annual averages have been calculated using the monthly data provided by the Oceanography Department of the Spanish Institute of Marine Research (which belongs to the Council of Scientific Research, CSIC). Figure 1 shows the results for the different maritime locations of the fishing grounds. These maritime locations range from 35°N to 45°N and from 8°W to 12°W for the Atlantic Ocean; and from 43°N to 45°N with a horizontal movement from 2°W to 8°W, which covers the entire Cantabrian coast. The data has been gathered for each grid of 1° each side, both vertically and horizontally. In order to carry out an analysis of the evolution of possible local warming in the study area, we have divided it up into three zones according to the different average temperature values observed:

from the Gulf of Cadiz (in the south of Spain) to the coast of Porto, from Porto to the boundary with the Cantabrian sea in Galicia, and for the rest of the Cantabrian sea. As can be observed in Figure 1, the hottest zone is Cadiz-Porto, which exceeds the average temperature of Porto-Galicia by approximately 2°C and the Cantabrian coast by approximately 1.5°C.

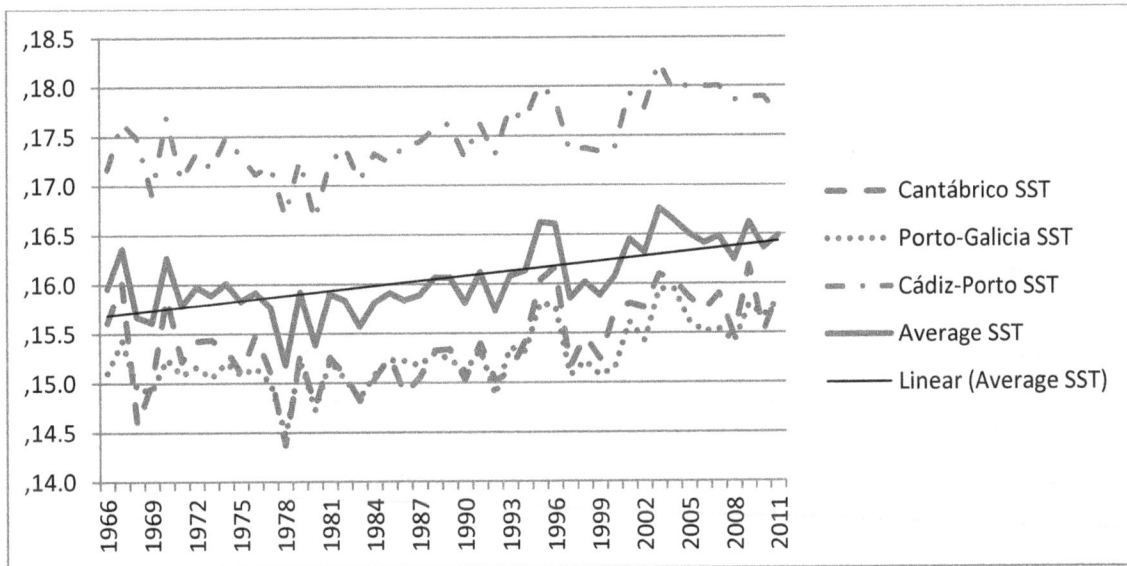

Source: Own compilation from Spanish Centre for Higher Scientific Research.

Figure 1. Sea surface temperature evolution (°C). 1987-2011

Although Figure 1 shows periods of average temperature increase (1968-1970, 1992-1996, 2000-2004) and periods where the temperature dropped (1975-1978, 1996-1998, 2003-2009), the overall trend shows an increase of the sea surface temperature for the period 1966-2011. On the other hand, in Table 1, some statistical measurements of the temperature series per maritime location are shown, this variable being highly disperse for the zone as a whole (0.35°C), greater in the Cantabrian zone (0.43°C). The minimum temperature value for the fishery overall was seen in 1978 (15.18°C), the maximum recorded in 2003 (16.76°C) And although the trend in the average temperature observed over the period 1966-2011 is upward, this increase was especially significant in the last thirty years of the period analysed: from the beginning of the decade of the 1980s. From that moment, the annual average sea surface temperature increased at a rate of 0.27°C per decade. Given that we do not know the foreseeable scenario with regard to warming for this fishing ground (the IPCC does not carry out forecasts at maritime zone level), this previous piece of data will be the one we use as a reference in the model applied, and we will assume that such a trend towards increases in SST will be maintained over the coming decades. In any case, we will devise an additional alternative scenario.

	Cantabrian zone SST	Porto-Galicia SST	Cádiz-Porto SST	Average SST
Standard deviation	0.431	0.321	0.362	0.351
Average	15.410	15.270	17.490	16.060
Variation coefficient	0.028	0.021	0.021	0.022
Minimum value	14.370	14.480	16.660	15.180
Maximum value	16.180	15.950	18.240	16.760

Source: Own compilation from Spanish Centre for Higher Scientific Research.

Table 1. Statistical indicators for SST

2.2. The sardine stock

Similarly to most pelagic species, the sardine lives in relatively shallow water, in areas of high primary productivity (phytoplankton and zooplankton), generally situated on the edges of anti-cyclone areas and where intense upwelling phenomena that bring nutrients to the surface occur. International Council for Exploration of the Sea (ICES) scientists estimate that the northern and southern limits of sardine distribution could be related to the average water temperatures, as they need to be in waters with temperatures of between ten and twenty degrees centigrade. Therefore, it is a fishery which is particularly sensitive to the effects of climate change, and this could result in either a drop in productivity or the movement of the biomass to cooler waters. Since 1980, the ICES, as the institution in charge of evaluating the situation of the European sardine, considers the management unit to be the zone which lies between the French and Spanish maritime border and the Strait of Gibraltar. Most of the catches of these species take place in waters of the Ibero-atlantic continental platform, in the zone known as sardine stock distribution area in ICES divisions VIIIc and IXa.

The EU has regulated this fishery solely through the establishment of minimum catch sizes since 1999 (set at 11 cm). The Spanish government has limited catches per vessel and day (7,500 kg for sizes larger than 15cm and 500 for sizes of between 11 and 15cm), and has established a temporary ban during which it is not possible to fish (from February to March, inclusive, each year). For their part, various regional governments have established a weekly two-day ban in order to regulate the fishing effort in this fishery [38].

	Biomass	Landings
	(tons)	(tons)
1978	328000	145609
1979	358000	157241
1980	456000	194802
1981	554000	216517

	Biomass	Landings
	(tons)	(tons)
1982	583000	206946
1983	506000	183837
1984	624000	206005
1985	723000	208439
1986	629000	187363
1987	548000	177696
1988	514000	161531
1989	528000	140961
1990	481000	149429
1991	465000	132587
1992	699000	130250
1993	875000	142495
1994	797000	136582
1995	811000	125280
1996	548000	116736
1997	466000	115814
1998	383000	108924
1999	345000	94091
2000	296000	85785
2001	408000	101957
2002	462000	99673
2003	436000	97831
2004	417000	98020
2005	386000	87023
2006	518000	96469
2007	481000	101464
2008	364000	87740
2009	276000	89571
2010	228000	80403
2011	224000	54857

Source: Own compilation from [35].

Table 2. The sardine biomass and landings in ICES Zones VIIIc and IXa. 1978-2011

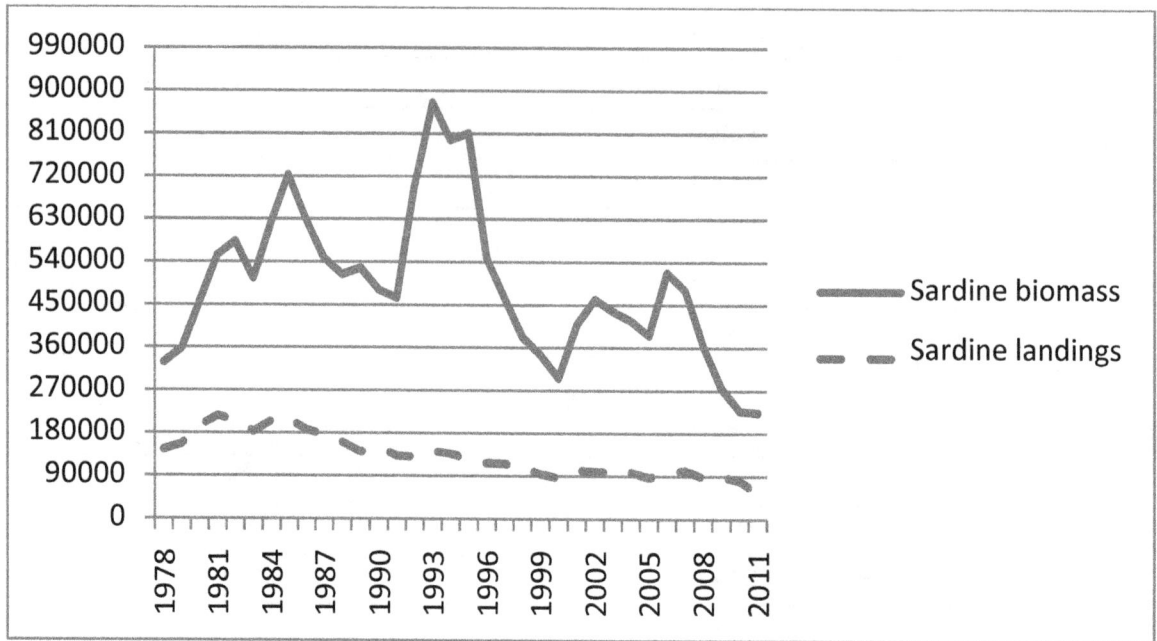

Figure 2. The sardine biomass (tons) and landings (tons) trends. 1978-2011

Table 2 shows the sardine biomass and landings data for the period 1978-2011 on the basis of information provided by ICES. In Figure 1, the evolution of both variables over said period can be better observed. As shown in this figure, the Ibero-atlantic sardine spawning stock biomass underwent signification oscillations in said period, with a notable increase of sardine biomass in the period 1990-1992, to then undergo a significant reduction until the beginning of 2000.

These phases in biomass evolution do not seem to be reflected with the same rhythm and intensity in the total catches of this species landed. If in the second half of the decade of the 1980s the figure stood at around 150,000 tonnes, after that it continued to drop gradually, especially in the last years of the study period and in spite of the estimated increase for sardine biomass in some specific years. This could be related to the progressive drop in the number of vessels which have fished in the fishery since the beginning of the 1990s.

On the other hand, being a pelagic species, the sardine is extremely sensitive to environmental changes in general, and to increases in ocean temperature in particular. In Figure 3 we can see the evolution of the spawning stock biomass and the average temperature of the water on the Cantabrian-north-west coast over the period 1984-2011.

It can be seen how, in general, after periods of slight increases in the sea surface temperature there is a drop in sardine biomass. This is especially significant in the periods 1994-1996 and 2003-2007. However, after periods when there was a drop in temperatures (1990-1992 and 1997-2003), an increase in the level of sardine biomass was seen. This can be proved from a statistical point of view. Table 3 shows the statistical correlation data regarding sardine

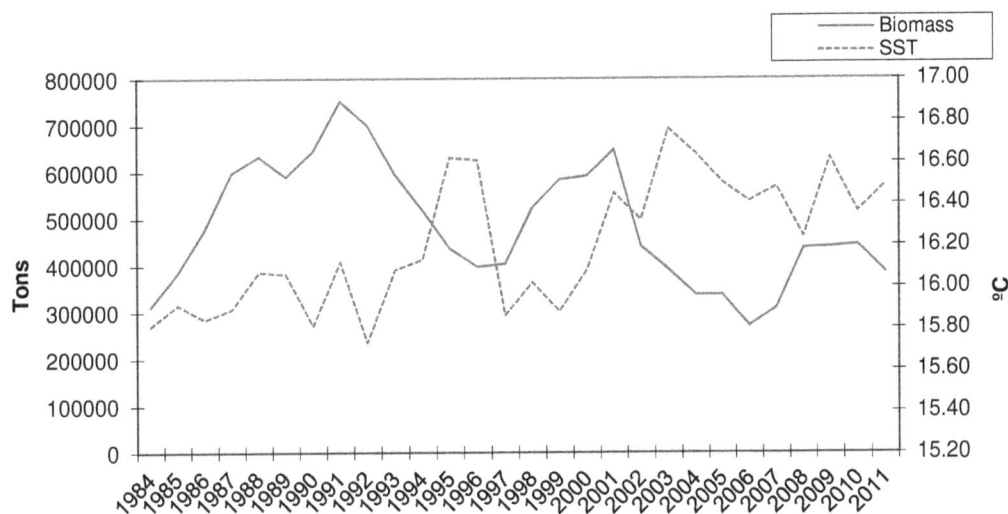

Source: Own compilation from [35].

Figure 3. The sardine biomass (tons) and sea surface temperature (°C). 1984-2011

biomass and average sea temperature. The negative temperature sign denotes that when variations in this variable occurred, the signs for variations in biomass went in the opposite direction.

	Sardine Biomass	Sardine biomass in t+1	Temperature (SST)
Sardine Biomass	1.0000		
Sardine biomass in t+1	0.86293	1.0000	
Temperature (SST)	-0.51841	-0.69284	1.0000

Source: Own compilation from Spanish Centre for Higher Scientific Research.

Table 3. Correlation matrix of variables

From all of the above, it can be deduced that oscillations in the sea temperature will impact on the natural productivity of this fish stock and, consequently, this will affect the generation of future economic yields through potential catches. On this basis, and in the face of an increase in the water temperature foreseeable for the northern hemisphere (as a consequence of global climate change; [6]), in this chapter we will make an estimation of its economic impact on the fishery. Prior to this, we will establish the zero scenario, our starting point, with which to make the comparison of possible future scenarios.

2.3. The economic data

Given that we do not have the economic data for the fleet whose target species is sardine as a whole, we are going to use a representative sample. The fleet sampled contributes around 18% of the total catches generated in the Ibero-atlantic sardine fishery in the period for which data was gathered (2001-2011), although this level of representation rises to 25% in the last three years of said period. Table 4 shows the evolution of the sampled vessels' landings and the income generated by their sale at market for the period we have been able to compile data for (2001-2011).

	Landings	Income	Average price
	(tons)	(euros 2011)	(euros/Kg)
2001	19 303.04	11 581824.0	0.60
2002	8 548.63	5 214664.3	0.61
2003	6 474.87	4 014419.4	0.62
2004	5 930.54	4 210683.4	0.71
2005	14 546.12	12 493 800.2	0.86
2006	11 491.22	10 661 267.3	0.93
2007	13 237.87	9 632 988.3	0.73
2008	19 668.66	13 715 626.4	0.70
2009	21 511.70	11 969 763.4	0.56
2010	20 227.14	10 951 081.8	0.54
2011	16 714.95	10 741 674.3	0.64

Source: Own compilation from www.pescadegalicia.com.

Table 4. Economic data: Income

With regard to the sample vessels' landings a downward trend can be observed in the first years of the decade 2000-2010, coinciding with years of low biomass, after which time the trend recovered gradually, except in the final years of the series when a drop in landings was observed once again after a new reduction in sardine biomass. In general, income followed a similar trend to landings. On the basis of both variables, the average sale price for this species for each year the period comprises can be obtained. As can be seen in Table 4, the average price in 2011 constant monetary units, on the other hand, displayed an upward trend until the middle of the decade, at which time it began to fall progressively.

With respect to the exploitation costs of this fishing activity, from [39] and the vessels sampled which provided this data, we have the information relating to the cost structure of this fleet in relation with the annual value of sardine landings. As can be seen in Table 5, the largest proportion of fleet costs goes towards paying crews' wages and salaries (this cost represents just over 55% of the average for the period overall), followed by the Gross Operating Surplus/ Mixed Income-shipowner/vessel owner income – (they make up approximately 17%), and vessel costs –mainly repairs and maintenance of the vessels themselves as well as fishing gear – (representing around 14%). Furthermore, the gradual increase in fuel costs is notable, and is especially significant in the last year of the period for which data is available.

From the information shown in the last two Tables, the average unit price and cost for the overall period have been estimated, and are shown in Table 6. The price of 10-year public bonds has been used as the social discount rate for the period considered. Against this background, we are in a position to establish the bio-economic model that will be used to estimate the effects deriving from changes in the sea surface temperature on the net profits of the fishery as a whole. We will use a period of approximately fifteen years, up until 2030, as a temporary horizon, as is proposed in the Climate Change Adaptation Plan drawn up by the former Ministry for the Environment.

%	2006	2007	2008	2009	2010	2011	Average 2006-11
Value of landings	100.00	100.00	100.00	100.00	100.00	100.00	100.00
Fuel costs	5.70	7.14	7.96	7.14	5.00	7.47	6.74
Other current costs	5.37	6.43	6.23	9.09	7.22	10.33	7.44
Vessel costs	16.11	14.29	13.15	12.34	16.67	10.55	13.85
Crew costs	53.69	53.93	57.09	54.87	56.39	55.60	55.26
Gross operating surplus/ Mixed income	19.13	18.21	15.57	16.56	14.72	16.04	16.71
Gross value added	72.82	72.14	72.66	71.43	71.11	71.65	71.97

Source: Own compilation from [39] and sample.

Table 5. Economic data: Cost Structure (%).

	Unit	Value of parameters
Unit Price of landings	€/ton	681.82
Unit cost of fishing	€/ton	554.56
Discount rate	%	5.00

Source: Own compilation from www.pescadegalicia.com, [39] and sample.

Table 6. Economic data: Unit values

2.4. Methodology

Bio-economic modelling allows introducing natural, environmental and institutional variables, in addition to the strictly economic, in one analytical body. For the specific case of fishing management problem, the aim is to control the stock of fish by limiting the catch (or effort) so that the stream of net benefits over time will be optimizedand tacking into account the dynamics of fishing stock. In this way, we can determine where society can invest (or divest) the marine resource and what should be the rate of abstraction allowing to maintain their sustainable exploitation.

The bio-economic problem is represented as follows:

$$Max_h \int_0^\infty \pi(X, h) e^{-\delta t} \, dt \tag{1}$$

$$s.a. \ \dot{X} = \frac{dX}{dt} = F(X) - h(t)$$

where denotes the net economic benefits obtained from fishing activity in each instant t, X represents the fish stock, h denotes the harvest rate, p the unit price of landings, c denotes the unit cost of fishing, δ the social discount rate and $F(.)$ represents the natural growth of fishing stock (the stock dynamics without considering the harvest). The optimization problem for regulator [40] consists of determining the feasible control, $h(t)=h^*(t)$ with $t \geq 0$, which optimises the net benefit stream while satisfying the problem's conditions in a global warming context.

Before solving the optimization problem (1), it is necessary to define previously the growth natural function $\dot{X}$. The stock dynamic is statistically tested on the basis of the data that exists on biomass and catches showed in Table 4; in addition, the sea surface temperature is included in this study case. Ordinary Least Squares (OLS) will be used to obtaining the parameters values. The Gordon-Schaefer expression is the function commonly used in the bio-economic literature [41-42], which is as follows incorporating the sea surface temperature (T):

$$\dot{X} = \alpha X_t + \beta X_t^2 + \gamma T_t - h_t \tag{2}$$

	$X_{t+1} + h_t = \alpha X_t + \beta X_t^2 + \gamma T_t$
α	2.057 (0.076)
β	-0.1048 E^{-5} (0.357)
γ	-6048.4 (0.712)
Jb	0.5582
Q-Stat	4.6335
LM (ARCH)	0.171
R^2	0.7528
R^2 adjusted	0.7250

Note: t-ratio between brackets. Jb is the Jarque-Bera statistic of the normality test; Q-Stat is the

Ljung-Box statistic used in the correlation test; LM (Lagrange multiplier) is the one used in the heteroscedasticity test.

Table 7. Econometric results for growth function of sardine stock

The equation (2) corresponds to the logistic model, where α, β and γ are parameters containing biological information on sardine stock, and T denotes the sea surface temperature. From data showed in Table 2, the results from the econometric estimation of (2) can be shown in Table 6 and so the stock dynamic is given by the following expression:

$$\dot{X} = 2.057\ X_t - 0.1048E^{-5}X_t^2 - 6048.4\ T_t - h_t \tag{3}$$

In this way, the *Hamiltonian* function in usual terms (current moment t) associated with problem (1) is given by the following expression [43]:

$$H\ (X,\ h,\ t;\ \mu) = (p - c)h_t + \mu\left(\alpha X_t - \beta X_t^2 - \gamma T_t - h_t\right) \tag{4}$$

where μ denotes the shadow price of the marine resource in current terms.

The conditions necessary [43] to solve the optimization problem are given by the following mathematical expressions:

$$\frac{dH\ (X,\ h,\ t;\ \mu)}{dh} = 0 ==> (p - c) - \mu = 0 \tag{5}$$

$$\dot{\mu} - \delta\mu = - \frac{dH\ (X,\ h,\ t;\ \mu)}{dX} = -\mu(\alpha - 2\beta X) \tag{6}$$

$$\frac{dX}{dt} = 0 ==> \alpha X - \beta X^2 - \gamma T - h = 0 \tag{7}$$

The condition (5) represents the known condition of economic efficiency. The condition (6) expresses the compensation which should exist in the optimum trajectory between the profit rate on the resource minus the social cost of not exploiting it (left side) and the total productivity of the same. Note that the natural resource is productive in two different ways: for its contribution to obtaining profits and for its contribution in the function itself of the natural growth of fish stock. And the condition (7) describes the evolution of the sardine biomass over the time.

By using expressions (5)-(7) the biomass and catch levels are obtained, $X^*(t)$ and $h^*(t)$, which will depend on the level of sea surface temperature. Once these levels are known, the losses or profits associated with the global warming scenario in relation with the fishery's present situation can be obtained through the net economic benefit function for this stady case.

3. Findings

In order to obtain numerical results for the impact on profits and given that the biomass levels in our model are going to depend on the sea surface temperature in the years to come, we are going to assume three possible scenarios for this latter variable. In the first scenario, we will assume that the oceanic temperature will increase at the same rhythm as the increase observed in the last decades (0.027°C per year). In the second scenario, we will assume that global warming intensifies and that that will affect the oceanic temperature, increasing to a greater extent than what has been seen to date (we will be estimating increases of 5% above the trend in the past decades). And in the third scenario, on the other hand, but in a context of rising temperatures, we will assume that global warming is being controlled to a greater extent (be it as a consequence of the current economic crisis or mitigation policies being carried out in the EU), and that this will give rise to a lower sea surface temperature in relation to that which would have occurred if the trend observed in the last decades continued (5% drop with regard to the past trend).

Table 8 summarises the three scenarios proposed. On the other hand, we will use up until the year 2030 as a temporary horizon, as is proposed in the Spanish government's Climate Change Adaptation Plan ([8-9]). In any case, we will extend it some years further in order to have a wider perspective of the evolution of the fishery's main indicators.

	Description	SST variation
Scenario 1	Trends last few decades	Increase of 0.027ºC yearly
Scenario 2	Acceleration in global warming and increasing in SST	Increase of 5% over trends last few decades
Scenario 3	Mitigation of global warming and in variation of SST	Decrease of 5% over trends last few decades

Table 8. Scenarios for trends in the sea surface temperature

Table 9 shows the scenarios forecast in relation with variations in sea surface temperatures. And the results obtained for the evolution of sardine biomass, catches and future profits the fleet could obtain are shown in Tables 10-12, respectively, for the different temperature variation scenarios contemplated.

	Scenario 1	Scenario 2	Scenario 3
2015	16.679	16.726	16.650
2016	16.706	16.754	16.677
2017	16.733	16.782	16.704
2018	16.76	16.811	16.731
2019	16.787	16.839	16.758
2020	16.814	16.868	16.785
2021	16.841	16.896	16.812
2022	16.868	16.924	16.839
2023	16.895	16.953	16.866
2024	16.922	16.981	16.893
2025	16.949	17.009	16.920
2026	16.976	17.038	16.947
2027	17.003	17.066	16.974

	Scenario 1	Scenario 2	Scenario 3
2028	17.030	17.094	17.001
2029	17.057	17.123	17.028
2030	17.084	17.151	17.055
2031	17.111	17.179	17.082
2032	17.138	17.208	17.109
2033	17.165	17.236	17.136
2034	17.192	17.264	17.163
2035	17.219	17.293	17.190
2036	17.246	17.321	17.217

Source: Own compilation.

Table 9. SST (°C) evolution under different scenarios

	Scenario 1	Scenario 2	Scenario 3
2015	223066.5	221837.425	224304.107
2016	217004.959	215510.916	218500.843
2017	205489.328	209431.150	212885.052
2018	205489.328	203543.027	207448.530
2019	200017.684	197880.085	202184.219
2020	194726.912	192393.630	197085.849
2021	189609.726	187114.034	192147.064
2022	184659.140	181996.034	187361.753
2023	179868.449	177070.778	182724.042
2024	175234.567	172292.792	178228.286
2025	170746.824	167690.392	173872.968
2026	166401.363	163224.606	169647.338
2027	162192.709	158920.764	165548.858
2028	158115.632	154742.593	161572.884
2029	155468.200	150713.953	157714.966
2030	150336.432	146801.044	153970.843
2031	146624.964	143026.362	150336.432
2032	143026.361	139358.379	146807.821
2033	139536.448	135818.318	143381.263
2034	136151.230	132376.743	140053.166

	Scenario 1	Scenario 2	Scenario 3
2035	132866.886	129053.689	136820.084
2036	129679.759	129680.000	133678.718

Source: Own compilation.

Table 10. Results for sardine biomass (tons) under different scenarios of SST

	Scenario 1	Scenario 2	Scenario 3
2015	136173.574	135682.327	136666.659
2016	133735.621	133134.565	134340.834
2017	131987.354	130633.823	132055.964
2018	128994.996	128184.904	129811.690
2019	126691.327	125782.099	127607.530
2020	124431.248	123429.341	125442.928
2021	122214.141	121121.508	123317.362
2022	120039.382	118862.051	121230.297
2023	117906.334	116645.972	119181.193
2024	115813.999	114476.667	117169.503
2025	113762.218	112349.624	115194.258
2026	111750.122	110267.480	113255.500
2027	109777.040	108226.212	111352.406
2028	107842.299	106228.127	109484.401
2029	105814.331	104269.509	107650.907
2030	104085.147	102352.38	105851.348
2031	102261.395	100473.302	104085.147
2032	100473.302	98634.044	102351.731
2033	98720.203	96831.428	100650.525
2034	97001.441	95067.000	98980.963
2035	95316.361	93337.816	97342.478
2036	93664.3171	91284.140	95734.509

Source: Own compilation.

Table 11. Results for the sardine catches (tons) under different scenarios of SST

	Scenario 1	Scenario 2	Scenario 3
2015	17329449.1	17266932.9	17392199.0
2016	17019195.1	16942704.8	17096214.5
2017	16796710.7	16624460.4	16805442.0
2018	16415903.1	16312810.9	16519835.6
2019	16122738.3	16007029.9	16239334.2
2020	15835120.6	15707617.9	15963867.1
2021	15552971.6	15413923.1	15693367.5
2022	15276211.7	15126384.6	15427767.6
2023	15004760.0	14844366.4	15166998.6
2024	14738489.5	14568300.6	14910990.9
2025	14477379.9	14297613.2	14659621.3
2026	14221320.6	14032639.5	14412895.0
2027	13970226.1	13772867.7	14170707.2
2028	13724010.9	13518591.4	13932984.9
2029	13465931.7	13269337.8	13699654.5
2030	13245875.9	13025363.8	13470642.6
2031	13013785.2	12786232.4	13245875.9
2032	12786232.4	12552168.5	13025281.2
2033	12563133.0	12322767.5	12808785.9
2034	12344403.3	12098226.4	12596317.3
2035	12129960.1	11878170.5	12387803.7
2036	11919721.0	11616819.7	12183173.6

Source: Own compilation.

Table 12. Results for the net profits (€) under different scenarios of SST

Starting with the evolution of sardine biomass (Table 10), as the sea surface temperature increases, the fish biomass decreases. The estimated biomass levels for the next twenty years are higher in the scenario of a lower increase than the one foreseen for temperature (scenario 3) and, on the other hand, they are lower in the temperature scenario with the greatest increase. This was to be expected, bearing in mind the relation observed between the sea temperature and the sardine biomass.

Along the same lines, the estimated catches evolve similarly to the fish biomass (Table 11). In particular, the fleet's potential catches dropped as the sea surface temperature rose over the period as a whole. However, the estimated catches are greater in the scenario where tempera-ture increases are lower (scenario 3) and lower when the temperatures are higher (scenario 2).

Lastly, and foreseeable given the forecast for the evolution of biomass and catches, the expected profits (Table 12) are higher in the scenario where the effects of climate change are mitigated (scenario 3), in which, with increasing sea surface temperatures, although lower than those proposed for the other scenarios, the profits would fall by 1.2% in the period (if the temperature were to drop 5% on the annual 0.027°C increase). In the remaining scenarios, the profits would fall more substantially: by approximately 1.25% (scenario 1) and 1.40% (scenario 2), approxi-mately, in the period, as the annual sea surface temperature increases.

4. Conclusions

The scientific evidence available to date leaves no doubt that climate, the different ecosystems and the planet, as we know it now, will not be the same in the future. Knowing the possible impacts that such change will bring is fundamental if mankind is to mitigate its effects and adapt to new global warming scenarios.

In this chapter, we have estimated the economic effects of climate change on one of the most relevant socioeconomic fisheries for the Spanish fisheries sector as a whole. This has been based on the effects of global warming on the oceanic temperature and its impacts on the natural productivity of the marine species. It has been observed that after slight temperature increases, a decrease in spawning biomass occurs, and vice versa. Specifically, and after estimating the correlation between the spawning stock biomass of sardine and the sea surface temperature, it has been identified that the temperature explains around 30% of the uncertainty associated with this fishery.

With regard to the estimation of the future impact, three possible temperature evolution scenarios have been taken into account: where the observed annual trend of a 0.027°C increase in the sea surface temperature is maintained; where global warming intensifies and therefore there is a greater increase in the sea surface temperature (increases of 5% above the previous trend); and where global warming is controlled and mitigated, implying a lower increase in the sea surface temperature with respect to the trend observed in recent decades (5% decrease).

The results show that as the sea surface temperature of the Ibero-atlantic fishing grounds increases, lower levels of biomass and catches are obtained, while the economic yields would be reduced. In particular, if the current sea surface temperature increase trend is maintained, annual profits will fall by 1.3% over the period analysed (2015-2036). If global warming intensifies and this were to generate an even greater increase in the water temperature of the fishing grounds, profits would drop by around 1.4% on average in each year of the period analysed. However, if palliative measures tending towards reducing warming were intro-duced, giving rise to a lower increase in the sea surface temperature, which could also derive from the current economic crisis and subsequent drop in production activity, profits would fall at an annual rate of approximately 1.2%.

In short, the trends are toward higher sea surface temperatures in these zones and, as a consequence, lower levels of spawning stock biomass of sardine. It has a direct negative impact on the catches and net benefits for the fishermen involved in the fishery: the higher sea temperature, the less catches and profits will be obtained.

The results obtained in this analysis are not intended to be definitive, however they do show from a scientific point of view how global warming can impact upon a specific economic activity whose main input is a natural renewable resource. The results obtained herein make it possible to identify and quantify, under specific assumptions regarding economic parameters and climate scenarios, the direction of the economic effects of global warming on one of the fisheries most sensitive to environmental shocks, as are the pelagic fisheries. These results demonstrate, once again, the need to increase palliative action and design strategies as to how best to adapt to the possible climate scenario, at both global as well as national or local level. Better knowledge of the effects of global warming will affect future decisions on the sustainable management of marine resources and the foreseeable reduction of the pressure they are under.

Acknowledgements

Special thanks to researchers from Spanish Centre for Higher Scientific Research (CSIC) and Spanish Oceanographic Institute (IEO) for their comments and suggestions. Financial aid from FEDER and Xunta de Galicia (GRC2014/022) is gratefully akcnoledged.

Author details

M. Dolores Garza-Gil*, Manuel Varela-Lafuente, Gonzalo Caballero-Míguez and Julia Torralba-Cano

*Address all correspondence to: dgarza@uvigo.es

Department of Applied Economics, University of Vigo, Vigo, Spain

References

[1] United Nations. Framework Convention of the United Nations on Climate Change. New York: United Nations Organization; 1992.

[2] IPCC. Climate change and biodiversity. Technical Issue V del IPCC. New York: UN; 2002.

[3] Admunson, R. The carbon budget in soils. Annual Review of Earth & Planetary Sciences 2001; 29: 535-562.

[4]   Dale V.H, Joyce L.A, McNulty S, Neilson R.P. The interplay between climate change, forest, and disturbances. Science of the Total Environment 2000; 262: 201-204.

[5]   Lugo, A.E. Biodiversity management in the 21 st. Century. Interciencia 2000; 26: 484-488.

[6]   IPCC. Fifth Assessment Report. WG1. New York: UN; 2013.

[7]   Garza-Gil M.D., Vázquez-Rodríguez M.X, Prada-Blanco A., Varela-Lafuente M. Global warming and its economic effects on the anchovy fishery and tourism sector in North-Western Spain. In Casalegno S. (Ed) Global Warming Impacts. Case studies on the economy, human health, and on urban and natural environments. Rijeka: In-Tech; 2011.

[8]   Ministerio de Medio Ambiente. Principales conclusiones de la evaluación preliminar de los impactos en España por efecto del cambio climático. Oficina Española de Cambio Climático (OECC). Madrid: Ministerio de Medio Ambiente; 2005.

[9]   Ministerio de Medio Ambiente. Plan Nacional de Adaptación al Cambio Climático. Madrid: Ministerio de Medio Ambiente; 2006.

[10] Pérez Muñuzuri V, Fernández Cañamero M, Gómez Gesteira J.L. Evidencias e impactos del cambio climático en Galicia. Samtiago de Compostela: Xunta de Galicia; 2009.

[11] Briones R, Garces L, Ahmed M. Climate change and small pelagic in developing Asia: the economic impact on fish producers and consumers. In Hannesson R, Barange M, Herrick S (Eds) Climate change and the economics of the world's fisheries. New York: Edward Elgar Publishing; 2006.

[12] Gallagher Ch. Variable abundance and fishery movements in New Zeland squid fisheries: Preliminary findings from global and regional investigations. Workshop on economic effects of climate change on fisheries. NHH: Bergen; 2005.

[13] Gómez Martín M. B. Weather, Climate and Tourism: a geographical perspective. Annals of Tourism Research 2005; 32(3): 571-591.

[14] Levitus S, Antonov J.I, Boyer T.P, Stephens C. Warming of the world ocean. Sciences 2000; 287: 2225-2229.

[15] Röckmann C. Rebuilding the Eastern Baltic cod stock under environmental change. Workshop on economic effects of climate change on fisheries. NHH: Bergen; 2005.

[16] Herrick S, Hill K, Reiss C. An optimal harvest policy for the recently renewed United States Pacific sardine fishery. In Hannesson R, Barange M, Herrick S (Eds) Climate change and the economics of the world's fisheries. New York: Edward Elgar Publishing; 2006.

[17] McGowan J.A, Cayan D.R, Dorman L.M. Climate-ocean variability and ecosystem response in North Pacific. Science 1998; 281: 201-217.

[18] Santos F. D, Miranda P. Alteraçoes climáticas em Portugal: escenarios, impactos e medidas de adaptaçao. Lisboa: Gradiva; 2006.

[19] Sissener E.H, Bjorndal T. Climate change and the migratory pattern for Norwegian spring-spawning herring: implications for management. Marine Policy 2005; 29: 299-309.

[20] Stenevik E.K, Sundby S. Impacts of climate change on commercial fish stocks in Norwegian waters. Marine Policy 2007; 31: 19-31.

[21] Arnason R. Global warming, small pelagic fisheries and risk. In Hannesson R, Barange M, Herrick S (Eds) Climate change and the economics of the world's fisheries. New York: Edward Elgar Publishing; 2006.

[22] Brander K. Climate change and fisheries management. In Quentin R, Hilborn R, Squires D, Tait M, Williams M (Eds) Handbook of Marine Fisheries Conservation and Management. New York: Oxford University Press; 2010.

[23] Hannesson R. Sharing the herring: fish migrations, strategic advantage and climate change. In Hannesson R, Barange M, Herrick S (Eds) Climate change and the economics of the world's fisheries. New York: Edward Elgar Publishing; 2006.

[24] Lam V, Cheung W, Swartz W, Sumaila R. Climate change impacts on fisheries in West Africa: Implications for economic, food and nutritional security. African Journal of Marine Science 2012; 34(1): 103-117.

[25] Rice J.C, García S.M. Fisheries, food security, climate change, and biodiversity: characteristics of the sector and perspectives on emerging issues. ICES Journal of Marine Sciences 2011; 68(6):1343-1353.

[26] ACIA (2004): Artic climate impacts assessment: Overview Report. Cambridge University Press, Cambridge.

[27] Guisande C, Cabanas, J.; Vergara, A.; Rivero, I. (2011). Effect of climate change on recruitment success of atlantic iberian sardine (Sardina pilchardus). Marine Ecology Progress Series 223, 234-250.

[28] Hollowed, A.; Barange M, Kim S, Loeng H, Peck M. Climate change effects on fish and fisheries: Forccasting impacts, assessing ecosystem responses, and evaluating management strategies. ICES Journal of Marine Science 2011; 68.

[29] Jennings S, Brander K. Predicting the effects of climate change on marine communities and the consequences for fisheries. Journal of Marine Systems 2010; 79(3-4): 418-426.

[30] Labarta U. Galicia Mariñeira: Historia económica y científica en estudio y explotación del mar en Galicia. Santiago de Compostela: Ed Universidad de Santiago de Compostela; 1979.

[31] Pestana G. Manncial Ibero-Atlántico de sardinha. Sua avaliaçao e medidas de gestao. Lisboa: Instituto Nacional de Investigaçao das Pescas; 1989.

[32] Domínguez M. La gestión de los recursos pesqueros en el marco de la PPC: Análisis de la cooperación. Vigo: Universidad de Vigo; 2003.

[33] MAGRAMA. La Agricultura, la Pesca y la Alimentación en España, Informe Anual. Madrid: Ministerio de Agricultura, Alimentación y Medio Ambiente; 2013.

[34] Ibermix. Identification and segmentation of mixed-species fisheries operating in the Atlantic Iberian Peninsula waters. Brussels: EC Project, FISH/2004/03-33; 2007.

[35] ICES. Report of the Working Group on the Assessment of Mackerel, Horse Mackerel, Sardine and Anchovy. Headquarters. Denmark: ICES; 2013.

[36] Porteiro C,; Cabanas J.M. Report of the Workshop on the Integration of Environmental Information into Fisheries Management. Strategies and Advice (anexe 6). Vigo: IEO; 2007.

[37] Rosón G. Índices climáticos y su impacto en la hidrografía y dinámica marina. Santiago de Compostela: Workshop Evidencias del cambio climático en Galicia; 2008.

[38] Garza-Gil M. D, Torralba-Cano J, Varela-Lafuente M. Estimating the economic effects of climate change on the European sardine fishery. Regional Environmental Change 2011; 11(1): 87-95.

[39] European Commission. Economic Performance of important segments fleets. Brussels: DG XIV; several years.

[40] Kamien M, Schwartz N. Dynamic Optimization. The Calculus of Variations and Optimal Control in Economics and Management. New York: North-Holland Ed; 1991.

[41] Clark C. W. Mathematical Bioeconomics-the Optimal Managemet of Renewable Resources. Sussex: J. Wiley and Sons; 1976.

[42] Clark C. W. Bioeconomic Modelling of Fisheries Management. New York: J. Wiley & Sons; 1985.

[43] Clark C.W, Munro G. R. The economics of fishing and modern capital theory: a simplified approach, Journal of Environmental Economics and Management 1975; 5(2): 96–106.

3

Influence of Greenhouse Gases to Global Warming on Account of Radiative Forcing

Akira Tomizuka

1. Introduction

Radiative forcing is a measure of the size of a greenhouse gas's contribution to global warming. Radiative forcing values are estimated by a numerical process using radiative transfer schemes for terrestrial radiation and data from general circulation models. However, the estimation is complex and difficult to understand for non-specialists, including researchers in other fields. Understanding the essence of the Earth system is important for correctly discussing global environmental issues. Accordingly, in this chapter, radiative forcing values are calculated from a simple, intuitive radiative transfer model using the absorption spectra of greenhouse gases and the Planck formula for terrestrial radiation. [1]

1.1. Radiative forcing

The global atmospheric carbon dioxide (CO_2) concentration has increased from about 278 ppmv in pre-industrial times (defined as 1750) to 390.5 ppmv in 2011. During the same period, the concentrations of methane (CH_4) and nitrous oxide (N_2O) have also increased from about 0.722 ppmv to 1.803 ppmv and about 0.270 ppmv to 0.324 ppmv, respectively (Table 1). [2]

Greenhouse gas	Concentration (ppmv)		Radiative forcing	Global warming potential	
	1750	2011	(W m^{-2})	20 yr	100 yr
Carbon dioxide	278	390.5	1.82 (1.63 to 2.01)	1	1
Methane	0.722	1.803	0.48 (0.43 to 0.53)	84	28
Nitrous oxide	0.270	0.324	0.17 (0.14 to 0.20)	264	265

Table 1. Concentrations in 1750 and 2011, radiative forcing, and global warming potential for each greenhouse gas. [2]

Radiative forcing is often referred to as an index of the size of a greenhouse gas's contribution to global warming. When the Earth system is at radiative equilibrium, the energy flux reaching the top of the Earth's atmosphere exactly balances with the outgoing energy flux from the Earth to outer space. However, increasing the concentration of greenhouse gases decreases the energy flux to outer space and changes the energy flux to the Earth into a surplus. Consequently, the Earth's surface temperature and atmospheric temperature rise, causing the outgoing flux to increase, and the Earth system shifts to a new equilibrium. Radiative forcing is defined as the imbalance of the energy flux density caused by these perturbations. The Intergovernmental Panel on Climate Change (IPCC) has estimated the following radiative forcing values due to increased greenhouse gas concentrations in 2011 relative to their pre-industrial levels: CO_2, 1.82 W m^{-2}; CH_4, 0.48 W m^{-2}; and N_2O, 0.17 W m^{-2}. The unit of radiative forcing is the same as that of energy flux density. The uncertainties in these values are all ±10% under 90% confidence intervals. While the magnitude of the positive radiative forcing of greenhouse gases is well understood, the effects of other atmospheric constituents such as aerosols are subject to considerable uncertainty.[3] IPCC estimated values are the sum of the contributions from direct effect (via emissions of gases) and several indirect effects (via atmospheric chemistry). Radiative forcing to be compared with calculated value in this chapter is the direct contribution: CO_2, 1.68 W m^{-2}; CH_4, 0.64 W m^{-2}; and N_2O, 0.17 W m^{-2}.

Global warming potential (GWP) is also used in comprehensive policies regarding the regulation of greenhouse gases. GWP is a measure of how much a given mass of a greenhouse gas contributes to global warming and is usually defined as the radiative forcing resulting from an instantaneous release of 1 kg of the greenhouse gas into the atmosphere relative to that of CO_2.[4] The GWP values for the next 20 and 100 years are given in Table 1.

Radiative forcing is estimated by a numerical process using radiative transfer schemes for terrestrial radiation and data from models referred to as general circulation models.[5] However, since the process is difficult to understand for non-specialists, including many citizens and researchers in other fields, they simply accept the results announced by the specialists. Yet, understanding the essence of the Earth system is important for correctly discussing global environmental issues. It is therefore necessary to create models such that anyone who has acquired basic scientific knowledge can intuitively understand the Earth system as well as the essence of the calculations based on the models.

In this chapter, radiative forcing values are calculated from a simple radiative transfer model using the absorption spectra of greenhouse gases and the Planck formula for terrestrial radiation. Furthermore, the GWPs of specific greenhouse gases are derived. Finally, the increase in the Earth's surface temperature due to radiative forcing is estimated.

1.2. How can radiative forcing be calculated?

The mean vertical temperature of the atmosphere results from the balance between heating and cooling. The Earth's surface and the troposphere are strongly coupled by convective heat transfer processes. At the surface, solar heating is balanced by convective transport of latent and sensible heat to the troposphere. In the troposphere, radiative cooling (infrared emission by molecules) is balanced by the release of latent heat via condensation and precipitation and by convective transport of sensible heat from the surface. This radiative-convective interaction leads to a roughly constant lapse rate in the troposphere. However, at the top of the troposphere (the tropopause), which is at an altitude of about 11 km above the Earth's surface, the temperature tends to become invariant with altitude.[6]

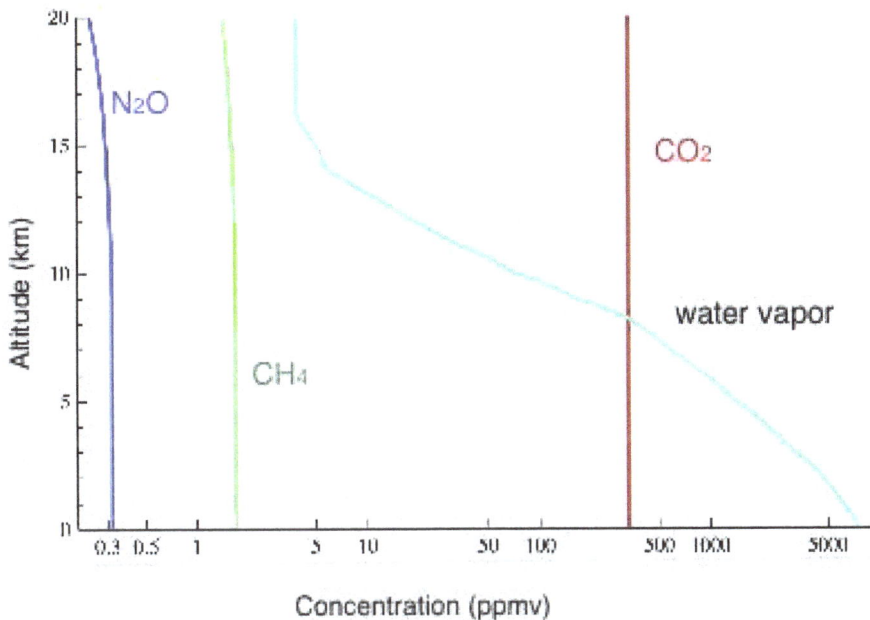

Figure 1. Vertical concentration profiles of four principal greenhouse gases. Concentrations are shown in units of ppmv. These data in the US Standard Atmosphere are obtained from the SpectralCalc website. [7]

The IPCC has defined radiative forcing as the change in net energy flux density at the tropopause. Figure 1 shows the altitude dependence of the concentrations of the four principal greenhouse gases. Water vapor is the most abundant and important greenhouse gas in the atmosphere. Nevertheless, it is excluded as an objective of the radiative forcing estimation since humans cannot directly control it. However, because of its strong absorption band, water vapor should be considered when calculating the radiative forcing of the other greenhouse gases.

The first step in building the model is to divide the atmosphere into appropriate layers, where the pressure, temperature, and concentration of each greenhouse gas are homogeneous within each layer. The concentrations of CO_2, CH_4, and N_2O are homogeneous between the Earth's surface and the tropopause, whereas the concentration of water vapor changes considerably. Moreover, the temperature of the troposphere decreases with altitude at a roughly constant lapse rate. For intervals with a thickness of 100 m, the change in the concentration of water vapor between the adjacent layers is around 5% and the change in temperature is 0.65 K.

Therefore, up to an altitude of 11 km, the atmosphere is divided into 110 layers, each with a thickness of 100 m. The bottom layer tangent to the surface is referred to as the zeroth layer for descriptive purposes.

The terrestrial energy flux emitted from the Earth's surface enters the zeroth layer. Some of the incident flux is absorbed by the molecules of greenhouse gases in the layer, and the remainder is transmitted. Subsequently, the molecules emit radiation flux both upward and downward. Some of the combined upward flux (transmitted flux and emitted upward flux) is again absorbed in the first layer, and the layer emits radiation.[8] For simplicity, the molecules emit radiation only once. The repetition of this simple radiative transfer process leads to the outgoing flux from the tropopause to outer space.

In the first step, the outgoing energy flux density $\bar{F}$ at the tropopause is calculated under the assumption that the concentrations of greenhouse gases in the atmosphere are equal to those in the pre-industrial era. In the next step, the flux density F_i is similarly calculated for an atmosphere in which the concentration of a specific greenhouse gas i increases up to its level in 2011, keeping everything else constant including the temperature.[4] The radiative forcing is obtained as $\Delta F_i = \bar{F} - F_i$ for the greenhouse gas i that changes in concentration.

In these processes, the effects of the near-infrared region of incident solar radiation are ignored. Also, the intensities of absorption and emission depend on only the respective number densities of the greenhouse gases in the layer. Pressure and temperature affect these intensities through only a change in number density. Thus, for the same concentration of a given greenhouse gas, the ratio of the absorption or emission of an arbitrary layer to that of the zeroth layer is equal to the ratio of the respective number densities for that gas. When concentration is dependent on altitude, as in the case of water vapor, the number density ratio is multiplied by the factor of altitude dependence. The relation between number density and altitude is discussed in the following section.

2. Mathematical relation of number density of gas molecules

Assume that the atmosphere within the troposphere is composed of an ideal gas with a density $\rho(z)$ at an altitude z and a mean molecular mass μ. The pressure $p(z)$ and the temperature $T(z)$ are described using the temperature in the zeroth layer T_0, the lapse rate in the troposphere Γ, and the universal gas constant R as follows:

$$p = \frac{R}{\mu}\rho T, \tag{1}$$

$$T = T_0 + \Gamma z. \tag{2}$$

Using the hydrostatic equation

$$\frac{dp}{dz} + \rho g = 0, \tag{3}$$

where g is the gravitational acceleration, we arrive at the following equation for the pressure:

$$\frac{dp}{dz} = -\frac{\mu g}{R(T_0 + \Gamma z)}p.$$

(4)

Equation (4) can be easily integrated from a height of 0 to z to obtain:

$$\frac{p}{p_0} = \exp\left[-\frac{\mu g}{R\Gamma}\ln(1+\Gamma z/T_0)\right].$$

(5)

Again, using Eq. (1), the number density ratio of the atmosphere $N(z)$ is as follows:

$$\frac{N}{N_0} = \frac{\rho}{\rho_0} = \frac{p}{p_0}\frac{T_0}{T_0+\Gamma z} = \frac{T_0}{T_0+\Gamma z}\exp\left[-\frac{\mu g}{R\Gamma}\ln(1+\Gamma z/T_0)\right].$$

(6)

The subscript 0 denotes the value for the zeroth layer. The number density of a well-mixed greenhouse gas $n(z)$ is related by the number density of the atmosphere:

$$n = C N,$$

(7)

where C is the concentration (volume mixing ratio) of the greenhouse gas. For the three gases (CO_2, CH_4, and N_2O) that have concentrations roughly independent of altitude, the ratio of number density is as follows:

$$\frac{n}{n_0} = \frac{T_0}{T_0+\Gamma z}\exp\left[-\frac{\mu g}{R\Gamma}\ln(1+\Gamma z/T_0)\right].$$

(8)

In the case of water vapor where the concentration changes with the altitude, Eq. (8) is multiplied by the factor of altitude dependence.

3. Simple radiative transfer

To obtain the change of outgoing energy flux density at the tropopause, it is necessary to calculate the radiative transfer of terrestrial radiation.[9] For a wavelength in the range between λ and $\lambda+d\lambda$, the net change in flux density passing through a layer with a thickness Δz is described as follows:

$$\Delta I\, d\lambda = -\sigma_a n \Delta z I\, d\lambda + \sigma_a n \Delta z B\, d\lambda,$$

(9)

where $I(\lambda, z)$ is the intensity, which is defined as the amount of radiant energy leaving a unit area of a body per unit time per unit spectral interval $d\lambda$; in other words, intensity is the energy flux density per unit spectral interval.

The first term on the right-hand side of Eq. (9) uses the Beer–Lambert law of absorption; $\sigma_a(\lambda)$ is the absorption cross section per unit greenhouse gas molecule and $n(z)$ is the number density of the greenhouse gas in the layer. The second term is blackbody radiation based on the Kirchhoff law, which states that emittance and absorption have identical values. $B(\lambda, z)$ is the intensity of the Planck blackbody function in the troposphere:

$$B = \frac{2\pi h c^2}{\lambda^5 \left(\exp\left[\dfrac{hc}{\lambda k (T_0 + \Gamma z)} \right] - 1 \right)}, \tag{10}$$

where h is the Planck constant, k is the Boltzmann constant, and c is the speed of light.

Scattering by molecules is ignored since the reference radiation is located in the infrared region. The absorption of the zeroth layer $a_0(\lambda)$ is defined as follows:

$$a_0 = \sigma_a n_0 \Delta z. \tag{11}$$

As a result, we can assume that the absorption of an arbitrary layer is described by $a_0(\lambda)$ and the ratio of number density is written as

$$\sigma_a n \Delta z = a_0 \frac{n}{n_0}. \tag{12}$$

Therefore, the net change in flux density is expressed as a function of altitude as follows:

$$\Delta I \, d\lambda = \frac{a_0 n}{n_0} (B - I) d\lambda = \frac{a_0 T_0}{T_0 + \Gamma z} \exp\left[-\frac{\mu g}{R\Gamma} \ln(1 + \Gamma z / T_0) \right] (B - I) d\lambda. \tag{13}$$

Taking into account the absorption by the four kinds of molecules, the actual change of the flux density passing through the layer is

$$\Delta I \, d\lambda = \left(w_0 f_w + \sum_{i=1}^{3} a_0^i \right) \frac{T_0}{T_0 + \Gamma z} \exp\left[-\frac{\mu g}{R\Gamma} \ln(1 + \Gamma z / T_0) \right] (B - I) d\lambda, \tag{14}$$

where a_0^i and w_0 represent the absorption spectra of anthropogenic greenhouse gas and that of water vapor of the zeroth layer, respectively. $f_w(z)$ is the factor of altitude dependence for water vapor, which is obtained by normalizing the concentration of an arbitrary layer by 7750 ppmv, which is the value for the zeroth layer.

The absorption spectra of greenhouse gases are discussed in the following section. The outgoing flux density at the tropopause with a wavelength from λ to $\lambda+d\lambda$ can be obtained by the repeated use of Eqs. (14) and (10) with an initial value of $I(\lambda, 0)=B(\lambda, 0)$. They are summed up for the reference range of wavelengths to give the total flux density. Of course, the total absorption through any layers must not be greater than one.

3.1. Estimation of radiative forcing

The calculations of radiative forcing require the absorption spectra of the four greenhouse gases for the zeroth layer in 1750.[10]

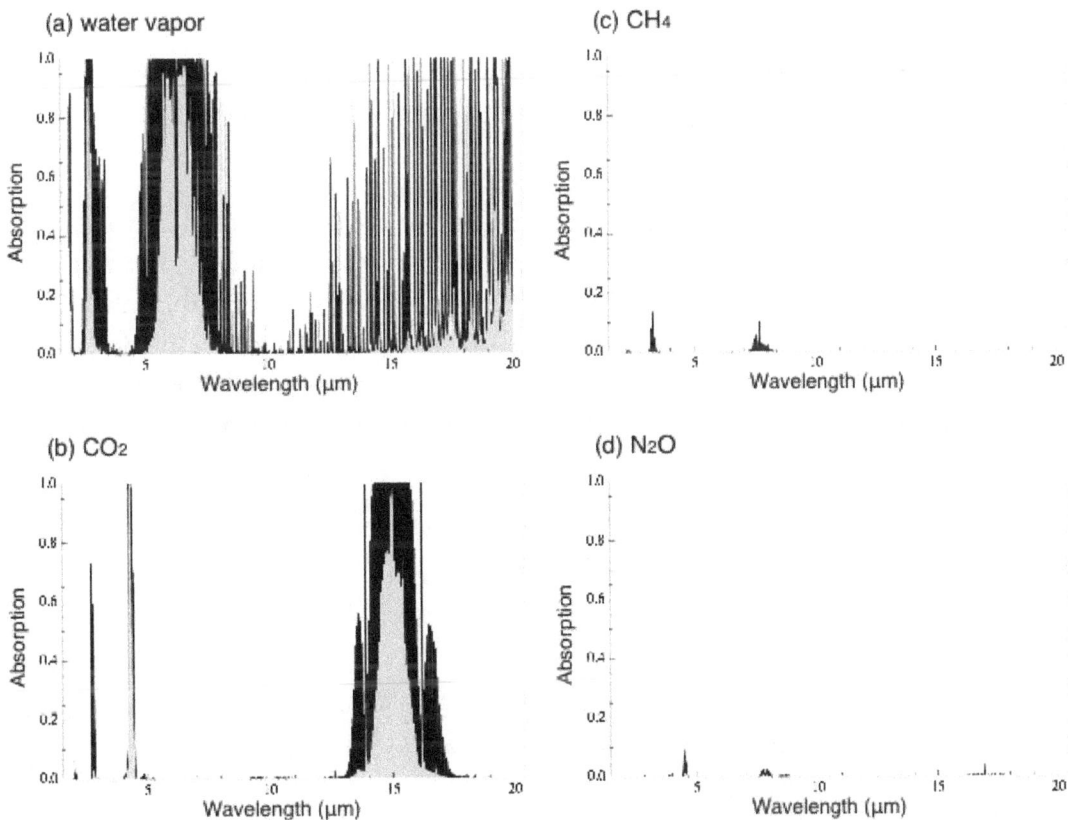

Figure 2. Absorption spectra of atmospheric greenhouse gases in 1750 for thickness of 100 m obtained from the SpectralCalc website: (a) water vapor, 7750 ppmv; (b) CO_2, 278 ppmv; (c) CH_4, 0.722 ppmv; and (d) N_2O, 0.270 ppmv. The concentration of water vapor is assumed to be the same as the present value. Absorption is the intensity ratio of absorbed radiation to total radiation incident on the zeroth layer.

Figure 2 shows the absorption spectra of the zeroth layer for 1750, as calculated using the SpectralCalc database. For selected gases, the SpectralCalc website provides transmittance spectra at an arbitrary pressure, temperature, and thickness of a gas layer; the concentration of the gas in the layer; and the range of wavelengths.[7] The SpectralCalc uses a line-by-line model called LINEPAK to accurately model molecular absorption line spectra.[11] These are based on the HITRAN database,[12] which is a compilation of spectroscopic parameters widely used to simulate the gases' transmission and emission of radiation into the atmosphere. The concentrations are 7750 ppmv for water vapor, 278 ppmv for CO_2, 0.722 ppmv for CH_4, and 0.270 ppmv for N_2O. The pre-industrial concentration of water vapor is assumed to be the same as the present value.

In the zeroth layer, the parameters are

$$T_0 = 288\,\text{K}, \quad p_0 = 1.013 \times 10^5\,\text{Pa}, \quad \Delta z = 100\,\text{m}, \quad \text{and} \quad \lambda = 1.85 - 20.0\,\mu\text{m}. \tag{15}$$

The unit spectral interval is set to 1 nm. The numerical values used for the calculations are as follows:

$$R = 8.3144\,\text{J}\,\text{K}^{-1}, \quad \mu = 28.964 \times 10^{-3}\,\text{kg}, \quad \Gamma = -6.5 \times 10^{-3}\,\text{K}\,\text{m}^{-1}, \quad g = 9.8\,\text{m}\,\text{s}^{-2},$$
$$h = 6.6261 \times 10^{-34}\,\text{J}\,\text{s}, \quad k = 1.3806 \times 10^{-23}\,\text{J}\,\text{K}^{-1}, \quad c = 2.9979 \times 10^8\,\text{m}\,\text{s}^{-1}. \tag{16}$$

Figure 3 shows the calculated outgoing spectra in the pre-industrial era at altitudes of 3 km, 6 km, 9 km, and 11 km (the tropopause). It is clear that terrestrial radiation is strongly absorbed at wavelengths of 5–8 μm and 13–17 μm. The former is due to water vapor and the latter is due to CO_2. In both these ranges, the absorption is nearly one. However, the outgoing flux in these ranges is not zero because of emission from the greenhouse gases. Moreover, the intensity of the flux decreases with altitude. The red area under the spectrum at 11 km, 191.60 W m^{-2}, corresponds to the total outgoing flux density $\bar{F}$.

The growth rates of greenhouse gas concentrations from the pre-industrial era to 2011 are given as follows: CO_2, 1.40; CH_4, 2.50; and N_2O, 1.20. For simplicity, we assume that the absorption spectrum in 2011 may be approximated by multiplying the spectrum in 1750 by the growth rate. The absorption values never exceed one.

A similar calculation in which the absorption spectrum of only 278 ppmv CO_2 is replaced with that of 390.5 ppmv CO_2 gives F_{CO2}=189.26 W m^{-2}. Therefore, the radiative forcing of CO_2 for the period between 1750 and 2011 is

$$\Delta F_{CO2} = \bar{F} - F_{CO2} = 2.34\,\text{W}\,\text{m}^{-2}. \tag{17}$$

Similarly, the outgoing flux density F_i for the spectrum of greenhouse gas i when its concentration in 1750 is replace with its concentration in 2011, is F_{CH4}=190.74 W m^{-2} and F_{N2O}=191.41

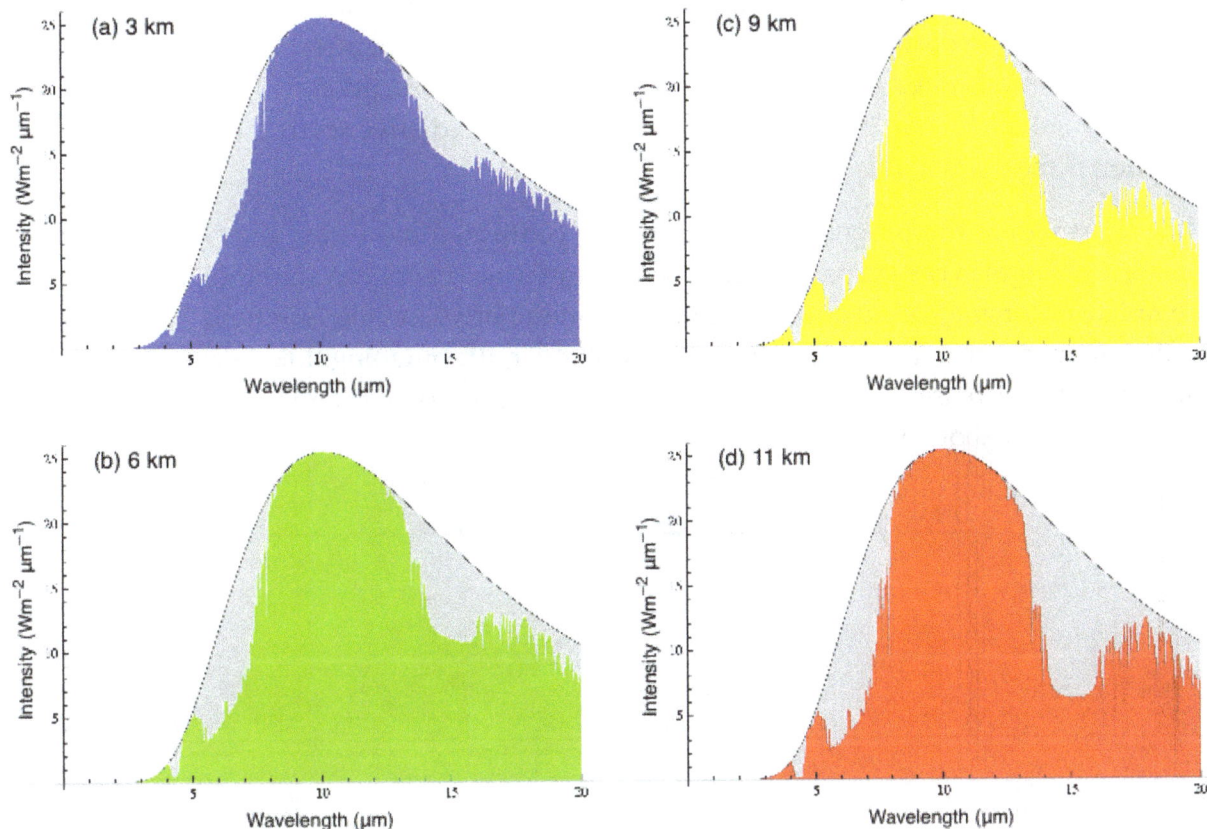

Figure 3. Outgoing terrestrial radiation spectrum at 3 km (blue), 6 km (green), 9 km (yellow), and 11 km (tropopause; red) in 1750, as calculated using the spectra shown in Figure 2.

W m^{-2}. The radiative forcing of each greenhouse gas is listed in Table 2. Despite of the simplicity of the radiative transfer model used here, the calculated values of ΔF are close to those in the IPCC estimations, but are relatively higher: CO_2, 39%; CH_4, 33%; and N_2O, 12%. The 2011 spectrum used in this calculation was approximated by multiplying the greenhouse gas's growth rate by the spectrum in 1750. Nevertheless, the differences between the approximated radiative transfer values and the precise values based on the real spectrum in 2011 in SpectralCalc are very small except for CO_2: CO_2, 12%; CH_4, 1.4%; and N_2O, 0.5%.

Greenhouse gas	Outgoing flux density		Radiative forcing	
	1750	2011	ΔF	IPCC*
Carbon dioxide		189.260	2.34	1.68
Methane	191.597	190.743	0.85	0.64
Nitrous oxide		191.407	0.19	0.17

* These estimates are direct contributions through the emissions of the gases.

Table 2. Calculated outgoing flux density in 1750 and 2011, radiative forcing ΔF, and IPCC estimates for each greenhouse gas.

Figure 4 shows the distribution of radiative forcing. We can see that the saturated absorption regions, 5–7 μm for water vapor and 16–18 μm for CO_2, do not affect the radiative forcing, whereas the unsaturated area does. Nevertheless, the CO_2 absorption in the range 9–11 μm is negligibly small (Figure 2(b)), contributing 10% of the radiative forcing because terrestrial radiation is maximal in this range.

The robustness of the proposed model must be confirmed by testing the sensitivity of the results to changes in certain parameters. A change in water vapor concentration does not affect the radiative forcing considerably. The increase in radiative forcing is around 1% even when water vapor concentration decreases by 5%. Moreover, the outgoing flux density decreases as the lapse rate increases. For a change of 10% in the lapse rate, the radiative forcing values change between about 3% and 6%.

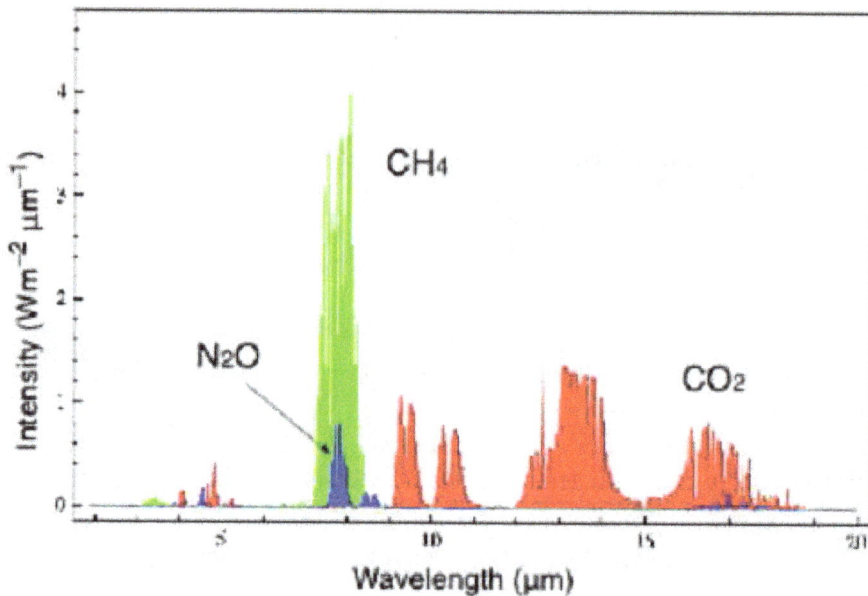

Figure 4. Radiative forcing spectra calculated with the precise absorption spectra: CO_2 (red), CH_4 (green), and N_2O (blue).

There are some sources of the uncertainties in this calculation. The first is assuming that the absorption spectra of greenhouse gases in each layer are proportional to only their number density. However, the absorption spectrum also changes with pressure and temperature in the layer: absorption line width broadening is caused by the thermal motion of the molecules and the collisions between them, both of which depend on pressure and temperature.[13,14]

This effect is captured by $a_0(\lambda)$ provided by SpectralCalc. However, the effects of the other layer are approximated as the effects of the zeroth layer for simplicity. These differences in the spectra in each layer will produce uncertainties in the outgoing fluxes.

The second source of uncertainty is ignoring the radiative transfer of incident solar radiation in the near-infrared region. The absorption and emission at 2.6–3.5 μm may not be negligible since the solar intensity in the region is 5–8 W m^{-2} μm^{-1} (the maximum intensity of the terrestrial radiation is about 25 W m^{-2} µm^{-1}, as shown in Figure 3). Therefore, the outgoing flux contains some uncertainties.

Figure 5 shows the calculated radiative forcing for each increasing greenhouse gas from the pre-industrial era to the present age. This result implies that each radiative forcing becomes slightly rounded due to saturation over the specific wavelength areas and it is linearly approximated by the gas's concentration: the radiative forcing per ppmv of CO_2 is 0.025 W m^{-2}; CH_4, 0.7 W m^{-2}, and N_2O, 3.5 W m^{-2}.

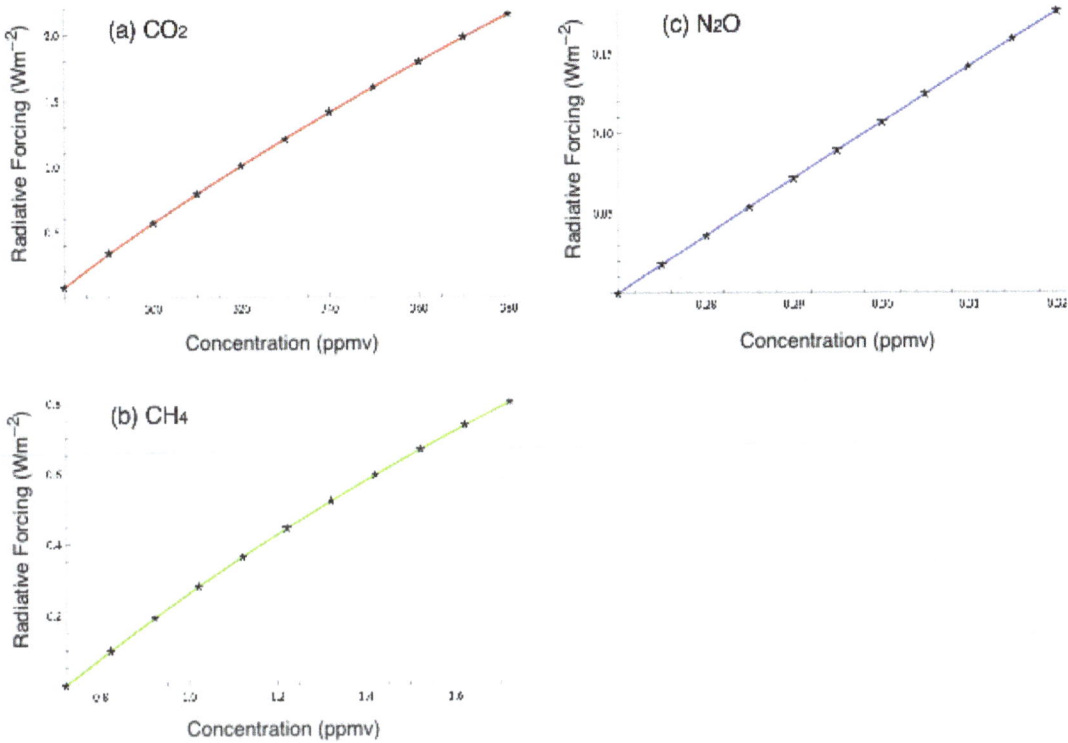

Figure 5. Calculated radiative forcings for increasing concentrations of greenhouse gases: (a) CO_2, for every 20 ppmv from 280 ppmv to 380 ppmv; (b) CH_4, for every 0.1 ppmv from 0.7 ppmv to 1.7 ppmv; (c) N_2O, for every 0.005 ppmv from 0.270 ppmv to 0.320 ppmv.

3.2. Estimation of global warming potential

GWP is defined as the radiative forcing per unit mass of released greenhouse gas i relative to that of CO_2. It is necessary to integrate the radiative forcing for a reference time since the number of gas molecules decreases with time. The Kyoto Protocol is based on GWPs from pulse emissions over a 100-year time frame. For simplicity, the GWP of greenhouse gas i for its 100-year effect is approximated as

$$GWP_i = \frac{(\Delta F / \Delta n)_i / M_i \times \int_0^{100} e^{-t/\tau_i} dt}{(\Delta F / \Delta n)_{CO2} / M_{CO2} \times \int_0^{100} e^{-t/\tau_{CO2}} dt}, \quad (18)$$

where $\Delta F / \Delta n$ is the radiative forcing per unit increased concentration and M is the molecular mass of the greenhouse gas. The mean residual life of a greenhouse gas over 100 years is given by

$$\int_0^{100} e^{-t/\tau} dt, \tag{19}$$

where τ is its lifetime.

The lifetimes of CH_4 and N_2O are 12.4 and 121 years, respectively, according to the IPCC Fifth Assessment Report. [2] For CO_2, the mean residual life is given as follows:

$$\int_0^{100} (0.217 + 0.259\, e^{-t/172.9} + 0.338\, e^{-t/18.51}) dt. \quad [15] \tag{20}$$

The GWPs of CH_4 and N_2O for the 100-year effect are calculated as 27.24 and 242.2, respectively, using ΔF as obtained in Section 3.1. They are listed in Table 3 along with the IPCC estimates. The GWP of CH_4 is very close to the IPCC estimate whereas the radiative forcing value is not.

Greenhouse gas	Δn (ppmv)	τ (years)	GWP	
			This work	IPCC
Carbon dioxide	112	*	1	1
Methane	1.08	12.4	27.2	28
Nitrous oxide	0.054	121	242	265

* For carbon dioxide, the IPCC uses the parameters in Eq. (20) which represent the mean residual life, without defining its lifetime. [15]

Table 3. Increased concentration Δn from 1750 to 2011, lifetime τ, calculated GWP, and IPCC estimate of GWP over an interval of 100 years for each greenhouse gas.

3.3. Radiative forcing and surface temperature

Finally, we estimate the increase in the Earth's temperature by using radiative forcing. In the pre-industrial era, the incoming solar radiation flux at the tropopause exactly balanced the outgoing solar radiation flux reflected by the Earth system and the outgoing terrestrial radiation.

Let the outgoing terrestrial radiation be approximated roughly as

$$\bar{F} = (1 - f)\sigma T_0^4, \tag{21}$$

where f is the mean absorption of the terrestrial radiation at the tropopause, T_0 is the surface temperature and σ is the Stephan–Boltzmann constant.

Increasing abundance of a greenhouse gas (e.g., CO_2) decreases the mean absorption by Δf.

$$F_{CO2} = (1 - f - \Delta f)\sigma T_0^4. \tag{22}$$

The perturbation Δf is related to radiative forcing by using Eqs. (21) and (22):

$$\Delta f = \frac{\Delta F_{CO2}}{\sigma T_0^4}. \tag{23}$$

Assuming Δf is maintained for some time, a new equilibrium state will develop in which the surface temperature has increased by ΔT. At the new radiative equilibrium, the outgoing terrestrial flux at the tropopause does not change if the incoming solar flux and the outgoing solar flux reflected by the Earth system are unchanged. Therefore,

$$(1 - f)\sigma T_0^4 = (1 - f - \Delta f)\sigma(T_0 + \Delta T)^4. \tag{24}$$

Ignoring the second-order and higher terms in Eq. (24), we obtain

$$\Delta T = \frac{\Delta f}{4(1 - f)}T_0. \tag{25}$$

Substituting Eqs. (21) and (23) into Eq. (25) gives the relationship between ΔT and ΔF_{CO2}:

$$\Delta T = \frac{\Delta F_{CO2}}{4\bar{F}}T_0. \tag{26}$$

By substituting the numerical values in Table 2, we can estimate that the Earth's surface temperature has risen by ΔT=0.88 K as a result of the CO_2 increase since 1750. Then, in the new equilibrium state, the surface temperature is $T_0'=T_0+\Delta T$=288.9 K and the lapse rate should be -6.6×10^{-3} K m^{-1} due to the temperature invariance at the tropopause. Next, the outgoing flux density at the tropopause with the surface temperature T_0' is calculated. This value is close to $\bar{F}$.

The radiative forcing of CO_2 is dominant, and the radiative forcings of other greenhouse gases nearly balance with negative radiative forcings (e.g., of aerosols). As a result, the influence of all radiative forcings is almost equivalent to the influence of CO_2.

Figure 6 shows the surface temperature increase due to CO_2 considering radiative forcing when the atmospheric CO_2 concentration increases to twice that in the pre-industrial era. In the future (when the CO_2 concentration exceeds 390 ppmv), the surface temperature increase per 20 ppmv CO_2 is about 0.1 K.

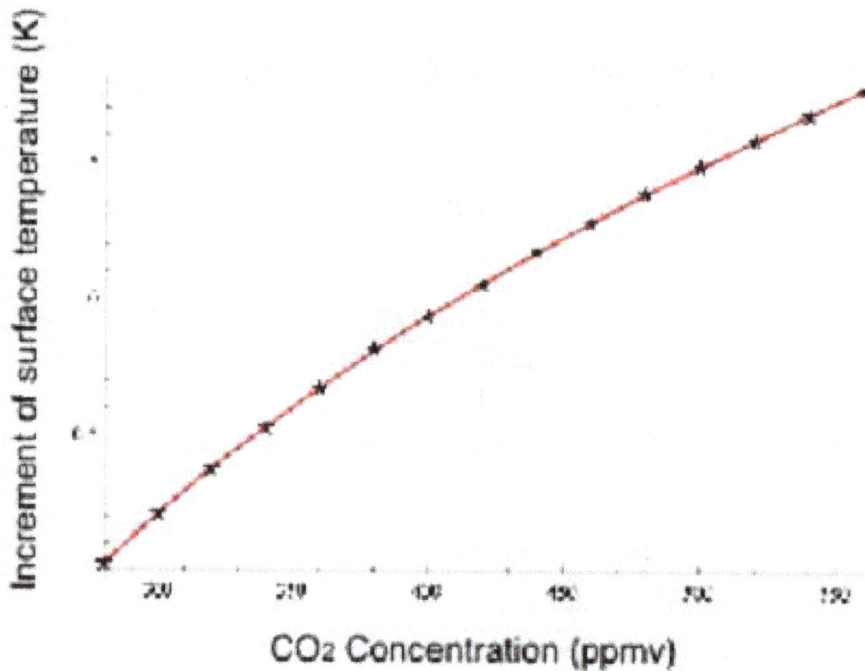

Figure 6. Surface temperature increase due to the radiative forcing of CO_2 when the atmospheric CO_2 concentration increases to twice that in the pre-industrial era.

4. Conclusions

The one-dimensional radiative equilibrium model can estimate the radiative forcing and global warming potential of anthropogenic greenhouse gases by using their absorption spectra at Earth's surface and number densities. Under some approximations, radiative forcing values are calculated higher than the IPCC estimates; CO_2, 39%; CH_4, 33%; and N_2O, 12%. Calculated values of global warming potential are very similar to IPCC estimates. Furthermore, increases in Earth's surface temperature due to radiative forcing can be easily estimated by the model.

Author details

Akira Tomizuka

Graduate School of Fisheries Science and Environmental Studies, Nagasaki University, Japan

References

[1] This chapter is based on the author's paper: A. Tomizuka, "Estimation of the power of greenhouse gases on the basis of absorption spectra," Am. J. Phys. 78, 359-366 (2010).

[2] "Climate Change 2013: The Physical Science Basis" (2014), http://www.ipcc.ch/report/ar5/wg1/.

[3] M. D. Mastrandrea and S. H. Schneider, "Resource Letter GW-2: Global Warming," Am. J. Phys. 76, 608-614 (2008).

[4] D. J. Jacob, *Introduction to Atomospheric Chemistry* (Princeton University Press, Princeton, NJ, 1999).

[5] G. Myhre, E. J. Highwood, K. P. Shine, and F. Stordal, "New estimates of radiative forcing due to well mixed greenhouse gases," Geophys. Res. Lett. 25, 2715-2718 (1998).

[6] *Radiative Forcing of Climate Change: Expanding the Concept and Addressing Uncertainties*, edited by D. J. Jacob, J. Antonio J. Busalacchi, and R. J. Serafin (National Academies Press, Washington, DC, 2005).

[7] "Spectral Calculator" (GATS Inc Atmospheric Science, Newport News, VA), http://spectralcalc.com/calc/spectralcalc.php.

[8] J. Harte, "Earth's Surface Temperature," in *Consider a Spherical Cow: a course in environmental problem solving* (University Science Books, California, 1988), pp. 160-171.

[9] F. W. Taylor, "Radiative Transfer," in *Elementary Climate Physics* (Oxford University Press, New York, 2005), pp. 67-103.

[10] J. Barrett, "Greenhouse molecules, their spectra and function in the atmosphere," Energy & Environment 16, 1037-1045 (2005).

[11] L. L. Gordley, B. T. Marshall, and D. A. Chu, "LINEPAK: Algorithm for Modeling Spectral Transmittance and Radiance," J. Quant. Spectrosc. Radiat. Transfer 52, 563-580 (1994).

[12] L. S. Rothman, "The HITRAN Database", 2009, http://www.cfa.harvard.edu/HITRAN/.

[13] R. H. Dicke, "The Effect of Collisions upon the Doppler Width of Spectral Lines," Phys. Rev. 89, 472-473 (1953).

[14] G. Visconti, "Molecular Spectra," in *Fundamentals of Physics and Chemistry of the Atmosphere* (Springer-Verlag, Berlin, 2001), pp. 324-328.

4

A Study on Assessment of Power Output by Integrating Wind Turbine and Photovoltaic Energy Sources with Futuristic Smart Buildings

Akira Nishimura and Mohan Kolhe

1. Introduction

Fossil fuel reserves are limited and intensive burning of hydro-carbon based fuel sources are impacting on global climate. In all over the world, there is continuous encouragement to increase the penetration of environment friendly energy sources for fulfilling growing energy demand and also to minimize the use of hydro-carbon based power plants. Renewable energy sources such as wind, photovoltaic (PV), solar thermal, geothermal, bio-energy are drawing attention from the world as alternative environment friendly energy sources. The energy density of these renewable energy sources is low. Most of them are dependent on nature and are found intermittent. It is very important to develop proper strategies to integrate these renewable energy sources into the power system network for fulfilling the energy demand. As cities around the world are experiencing exponential growth and there is urgent need to ensure that cities should expand sustainably, operate efficiently, and maintain a high quality life of residents. One of a city's most important critical infrastructures of a city is reliable power supply network. The smart city is an effective way to integrate renewable energy sources into the existing energy system network. In Japan, some demonstration projects of smart city are under contemplation [1]. In China, Tianjin City is being rebuilt as an ecological city by project in collaboration with a Singapore company [2]. Such type of trend will continue and in near future many cities will be rebuilt as smart city.

In a smart city planning, it is very important to consider the future growth of building integrated environment friendly energy systems. Energy output from intermittent renewable energy sources in the built environment depends on the availability of natural resources (e.g. wind speed, solar radiation, etc.) in the urban area. In the built environment, it will be challenging to integrate intelligently renewable energy sources and distributed generators as the existing building infrastructure are not designed to integrate them into the power system infrastructure. In future smart city planning, it is very important to consider proper intelligent integration of renewable energy sources into the built environment. A smart city development and deployment of building integrated renewable energy system has to harmonize with expansion of the combined heat and power system infrastructure, the information and communication infrastructure, the transport infrastructure, and with integration of new secure monitoring and control applications [3].

This chapter intends to propose a smart city for utilizing renewable energy sources as much as possible. For example, the city consists of many buildings and the buildings are thought to be an obstacle to natural wind flow. If the wind movements through building layouts are controlled, then there is possibility that the wind can be utilized for power generation through wind turbine. In addition, solar power can be utilized by installing solar panels on the roof and/or side wall of the buildings. The proper building design can help to utilize the available solar radiation for generating power and heat through solar energy systems. For designing a building in smart city, building dimensions and layouts are very important for effective utilization of wind speed and solar radiation. It has been observed that, there is very limited research and project works, which are investigating these issues. This chapter is providing study on large scale power generation by wind turbine and PV systems integrated with building and in the city. In this study, the horizontal axis wind turbine is considered for integrating with city infrastructure and the output of commercial horizontal axis wind turbine is much larger compared to that of the commercial vertical axis wind turbine. Additionally, the building integrated with PV systems is considered. The proposed schematic of smart buildings integrated with wind turbine and PV system is given in Figure 1.

In a smart city, it is very important to analyze the wind speed and their directions flow. The analysis of wind flows around the buildings is done through numerical simulations and most of them are using turbulent model such as standard k-ε, LES (Large Eddy Simulation), and DNS (Direct Numerical Simulation) [4-10]. Also, wind tunnel experiments on wind flows around a building have been discussed in references [11-14]. Although these works have investigated wind velocity profile around various building models under different conditions [4-14], but there has not been any report/work in which building sizes and layouts are considered in order to utilize wind blowing around the building for power generation from wind turbine in built environment. To realize wind energy utilization in the built environment, it is important to conduct feasibility study on power generation from wind turbine under the actual wind speed conditions. Although the power generation performance of a wind turbine has been predicted using frequency distribution of wind speed and wind direction, in most of the studies, the proper planning of wind turbine by utilizing the wind movements through building sizes and layouts and PV system in the built environment in a smart city are not considered [15-24]. Additionally, it is observed that there are very few studies that investigate the power output of building integrated PV system with wind turbine under actual meteorological conditions considering city layouts. Moreover, it is important to examine/analyze the

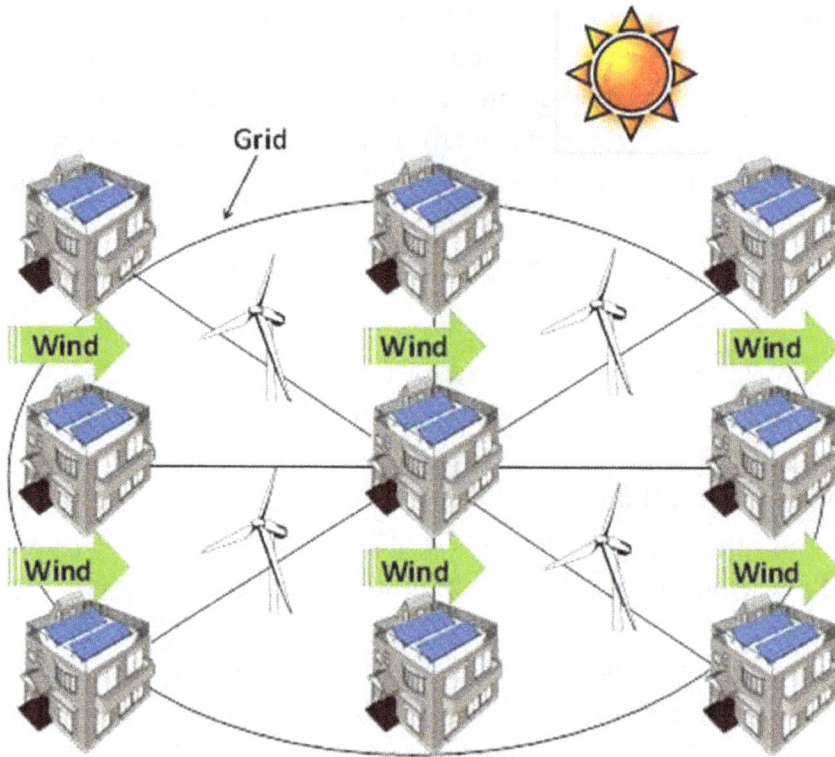

Figure 1. Image of smart buildings integrated with wind turbine and PV system proposed by this study

feasibility of installing wind turbines in planed building models and solar PV electricity generation and their role on meeting electrical energy demand of the city.

In this study, building layouts are considered for producing higher wind turbine power output in built environment [25, 26]. The configuration of building layouts like nozzle, as shown in Figure 2, is proposed and investigated to obtain the tapered wind flow through the buildings. Two buildings are configured as a nozzle (Figure 1) and the building size is taken 10 m width, 40 m length, and 40 m height. The representative length of this model L, which is a hydraulic diameter of horizontal cross area, is 16 m. Other dimensions (e.g. angle between two buildings, i.e. 90 degree, distance between two buildings 40 m, etc.) of the building layouts are given in Figure 2. In a city planning, there will be several buildings, but this study considers only two buildings. In future work, multi building layouts will be considered for wind speed distribution in the downstream. The results of numerical simulation on wind flows around buildings have been carried out by the turbulent model such as k-ε model. The wind speed distributions across the buildings according to the proposed building layouts are investigated. Moreover, the wind speed distribution under the various wind velocities and directions at inlet of the building model is investigated in order to simulate the actual meteorological conditions. The wind speed data base of the Japan Meteorological Agency [27] and the power curve characteristics of commercial wind turbine are utilized for evaluating wind speed profile across the buildings and for finding the electrical energy output from a wind turbine. In addition, this study presents the investigation results on the optimum installing procedure of solar PV panel on the roof of a building under the actual meteorological conditions in order to obtain higher

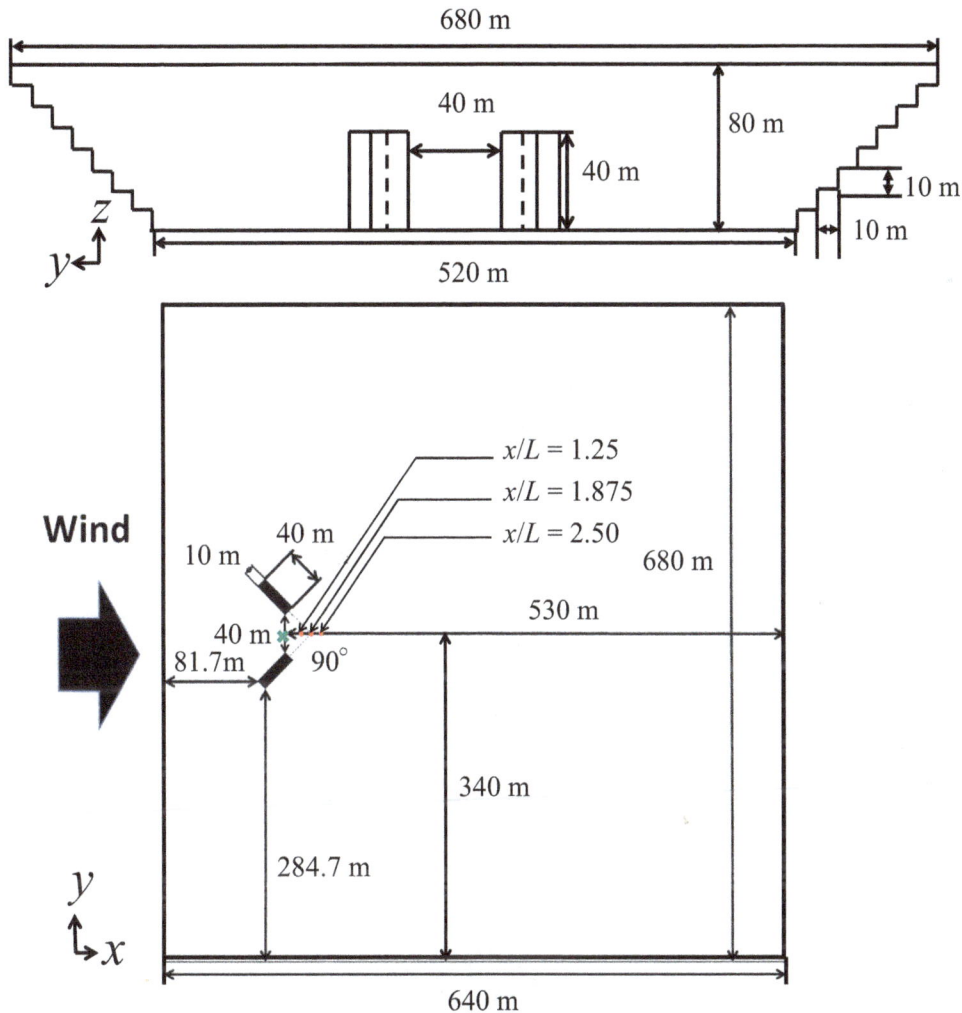

Figure 2. Building layouts for wind speed profile and wind power generation

power output from PV system. This study also evaluates the power generation characteristics of combined system including wind turbine and PV under the actual meteorological conditions as well as energy supply adaptability to the energy demand of a building. The change in power energy of wind turbine and PV system with time is investigated comparing with real time energy demand. It is very important to consider these aspects while designing a smart city and its infrastructure. The real time power generation form the energy sources located in the built environment and the demand characteristics are going to be very useful for designing and planning distributed smart power system network infrastructure. The significant point is to fulfill the built environment energy demand from the renewable energies through daily and seasonal variations and these analyses are presented in this chapter. It is assumed that proposed building models will be located in the actual city in Japan and local energy supply and demands are also discussed. The study, which is presented in this chapter, may be very useful for city planner for finding proper locations/layouts of the buildings for effective utilization of wind and solar energy resources in the built environment. It can suggest new concepts in order to construct/develop a smart city, which can help/contribute in reducing green house gas emissions.

2. Building model analysis in built environment

2.1. Simulation of wind speed distribution due to building sizes and layouts in built environment

In this section, a commercial CFD software CFD-ACE+(WAVE FRONT) is adopted for numerical simulation of wind speed distribution in the built environment. This CFD software has many simulation code/tools for solving the multi-dimensional fluid dynamics. The validation of the simulation procedure of this CFD software has been well established [28-33]. The standard k-ε model is adopted in this study. In the CFD software, the continuity equation is given by [34, 35]:

$$\frac{\partial \rho}{\partial t} + \nabla\left(\rho \vec{V}\right) = 0 \tag{1}$$

where ρ is density [in kg/m³], t is time [in s] and $\vec{V}$ is velocity vector [in m/s].

The momentum equation is given by [34, 35]:

$$\frac{\partial u_j}{\partial t} + \nabla\left(u_j \vec{V}\right) = -\frac{1}{\rho}\frac{\partial p}{\partial x_j} + \nabla\left(v_{eff}\nabla u_j\right) \tag{2}$$

$$v_{eff} = v + v_t \tag{3}$$

where u_j is velocity [in m/s] at j component of coordinate system, p is pressure [in Pa], v_{eff} is effective viscosity coefficient [in m²/s], v is kinematic viscosity coefficient [in m²/s] and v_t is eddy viscosity coefficient [in m²/s].

In the CFD software, the standard k-ε model is given by [34, 35]:

$$v_t = \frac{C_\mu k^2}{\varepsilon} \tag{4}$$

$$\frac{\partial k}{\partial t} + \frac{\partial}{\partial x_j}\left(u_j k\right) = S - \varepsilon + \frac{\partial}{\partial x_j}\left[\left(v + \frac{v_t}{\sigma_k}\right)\frac{\partial k}{\partial x_j}\right] \tag{5}$$

$$\frac{\partial \varepsilon}{\partial t} + \frac{\partial}{\partial x_j}\left(u_j \varepsilon\right) = C_{\varepsilon 1}\frac{S\varepsilon}{k} - C_{\varepsilon 2}\frac{\varepsilon^2}{k} + \frac{\partial}{\partial x_j}\left[\left(\nu + \frac{\nu_t}{\sigma_\varepsilon}\right)\frac{\partial \varepsilon}{\partial x_j}\right]$$ (6)

$$S = \nu_t\left(\frac{\partial u_i}{\partial x_j} + \frac{\partial u_j}{\partial x_i} - \frac{2}{3}\frac{\partial u_m}{\partial x_m}\delta_{ij}\right)\frac{\partial u_i}{\partial x_j} - \frac{2}{3}k\frac{\partial u_m}{\partial x_m}$$ (7)

where k is turbulent energy [in m²/s²], ε is dissipation rate [in m²/s²], δ_{ij} is Kronecker delta, C_μ is 0.09, $C\varepsilon_1$ is 1.44, $C\varepsilon_2$ is 1.92, σ_k is 1.0, $\sigma\varepsilon$ is 1.3. Regarding x_i, x_j, and x_m which represent components of coordinate system [in m], $x_1=x$, $x_2=y$, $x_3=z$. Regarding u_i, u_j, and u_m which represent velocities [in m/s], $u_1=U$, $u_2=V$, $u_3=W$. U, V, W are the velocity components of coordinate system, x, y, z, respectively.

Density of wind at inlet	1.166 kg/m³
Temperature of wind at inlet	293 K
Pressure of wind at inlet	0.1 MPa
Kinematic viscosity coefficient of wind	1.56×10^{-5} m²/s
Wind speed at inlet	$U = U_0\times(z/30)^{0.25}$ m/s ($U_0 = 3.00 - 12.00$ m/s)
Slip on side wall of building	$V = (0.41\times|l|)^{0.25}U$
Turbulent flow model	Standard k-ε model
Turbulent energy	0.025 m²/s²
Dissipation rate	$(1.58\times10^{-3})/z$ m²/s²
Calculation number	10000
Residue of each parameter	$<1.0\times10^{-5}$
Calculation state	Steady state

Table 1. Simulation condition of wind speed distribution around buildings

Table 1 lists the simulation parameters of wind speed distribution around buildings. The simulation model is already shown in Figure 2. Numerical simulation has been carried out under steady state by standard k-ε model and the calculation number is set 10000. This calculation number should be appropriate because the residue of each parameter under each numerical simulation condition keeps a stable low value after 500 times calculation. Wind speed at inlet of the model is set by the following equation:

$$U = U_0 \left(\frac{z}{30} \right)^{0.25} \tag{8}$$

where U is the wind speed [in m/s] in x direction, U_0 is the initial wind speed [in m/s] at z=30 m which is changed from 3.0 m/s to 12.0 m/s, z is height [in m]. U_0=10.0 m/s is the rated wind speed of AEOLOS (AEOLOS: wind turbine manufacture) wind turbine of 50 kW class [36]. In this equation, it is assumed that U equals to U_0 at z=30 m which is the hub height of wind turbine when the wind reaches to the building.

In this study, the wind speed data for Tsu city in Japan are used from the Japan Meteorological Agency [37] for five years (from 2007 to 2011). The buildings locations are considered as a nozzle, therefore the wind inflow direction is important for obtaining the wind blowing through the buildings. The layouts of the buildings are decided based on the wind speed direction. The wind speed directions and building layouts are given in Figure 3. If the main wind direction is North (N), the wind from North-West (NW), North-North-West (NNW), North-North-East (NNE) and North-East (NE) including North can be utilized for blowing the wind among the buildings through proposed nozzle. Assuming the symmetry to the main wind direction, the wind speed distributions around the buildings for the in-flow angles β of 22.5 degree and 45 degree are simulated to evaluate the effect of four angular inflows on the wind speed distribution. Wind speed at the inlet of the model is set by Eqs. (9) and (10), when the effect of inflow angle is considered.

$$U = \cos \beta \times U_0 \left(\frac{z}{30} \right)^{0.25} \tag{9}$$

$$V = \sin \beta \times U_0 \left(\frac{z}{30} \right)^{0.25} \tag{10}$$

where V is the wind speed [in m/s] in y direction, β is in-flow angle.

The top layer of the model has been considered free (without any disturbance). The slip on side wall of building is set by the following equation:

$$V = \left(0.41 \times |l| \right)^{0.25} U \tag{11}$$

where $0.41 \times |l|$ is the mixing length, 0.41 is Karman coefficient, l is distance from wall of building.

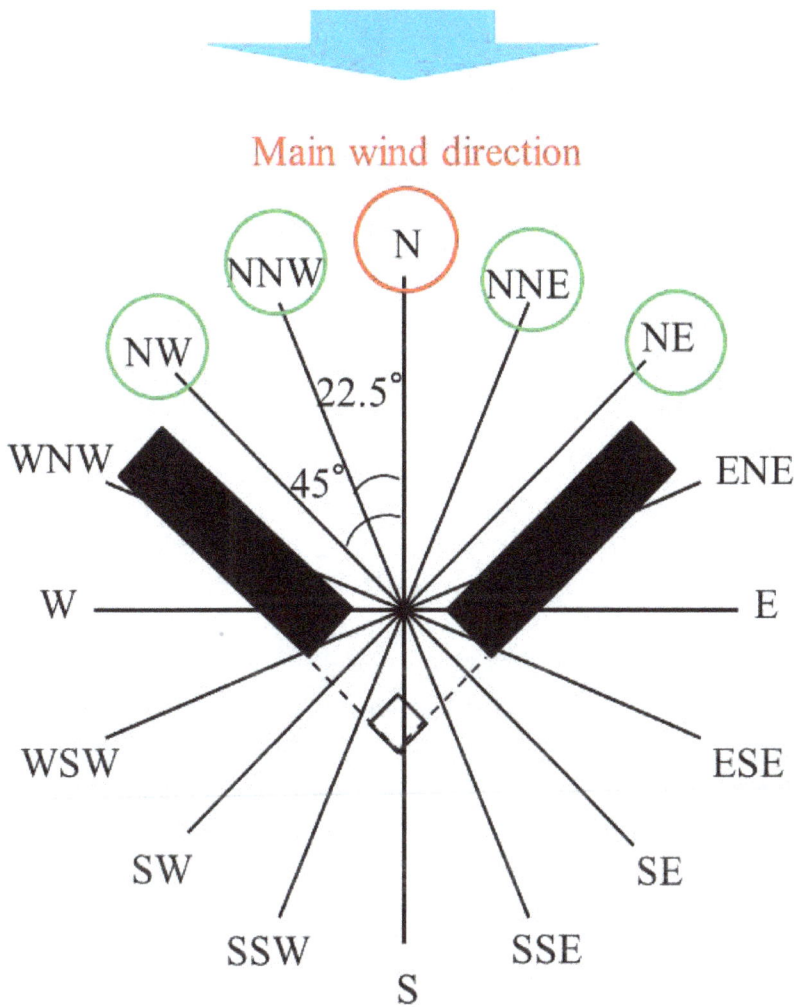

Figure 3. Wind speed directions and layouts of buildings

2.2. Concept for building size setting in built environment

It is assumed that buildings are multi storied apartments. According to the statistical data collected by ministry of internal affairs and communications in Japan [38], the average floor area of dwelling in Japan is about 100 m² per household. The height of one floor is assumed as 4 m. Assuming that four households stay per floor, the floor space is 400 m². In this study, for simulating a nozzle by buildings orientation, the width and depth of building are set at 10 m and 40 m, respectively. The height of building should be set over the height of wind turbine if we request an accelerated wind by blowing through buildings. In this study, the real commercial wind turbine is considered for estimating the power generated by wind speed distribution. AEOLOS wind turbine of 50 kW class [36] is used. Table 2 provides the specification of the wind turbine. The hub height and rotor radius of this turbine are 30 m and 9 m, respectively, resulting that the height of building of 40 m is almost same as axis height of wind turbine.

Rated power	50 kW
Start wind speed	3 m/s
Cut-in wind speed	3 m/s
Cut-out speed	25 m/s
Rotor diameter	18 m
Rotor speed	60 rpm
Hub height	30 m

Table 2. Specifications of wind turbine

2.3. Power generation estimation from built environment located wind turbine

The wind in back area of the building is considered to be available for power generation from wind turbine, as the wind would be accelerated by blowing through nozzle created by buildings layouts/orientations. Three points at the back of buildings, which are apart from the buildings by 20, 30, 40 m (x/L=1.25, 1.875, 2.50), are assumed as installation points of the wind turbine. The wind speed for calculating the power generated by wind turbine is obtained on 1049 points located in the area where the rotor of wind turbine rotates, i.e., the swept rotor area. The wind speed at each point on the swept rotor area is considered average wind speed in a local area of 0.5 m × 0.5 m. By considering the wind speed distribution of this local wind speed, the wind energy can be calculated. Average wind speed to x axis direction is estimated by using the following equation:

$$U_{ave} = \left(\frac{2Q_x}{N\rho A} \right)^{1/3} \tag{12}$$

where U_{ave} is the average wind speed [in m/s] to x axis direction, Q_x is the summation of wind energy [in W] to x axis direction on each point for calculating wind speed distribution in the swept rotor area A (=1049 points), ρ is the density [in kg/m³] of wind, A is the swept rotor area [in m²]. Wind energy at each point on the swept rotor area is calculated by the following equation:

$$Q_x = \sum_{i=1}^{1049} Q_{x,i} = \sum_{i=1}^{1049} \left(\frac{1}{2} \rho A_i U_i^3 \right) \tag{13}$$

where $Q_{x,i}$ is the wind energy [in W] to x axis direction at each point, A_i is the area of each point [in m²] which is equal to 0.5 m × 0.5 m, U_i is the wind speed [in m/s] to x axis direction at each point for calculating wind energy. V_{ave} which is the average wind speed [in m/s] to y axis direction is estimated by the same calculation way of U_{ave}. The average wind speed [in m/s] to horizontal surface of the swept rotor area $U_{h,ave}$ is calculated by the following equation:

$$U_{h,ave} = \sqrt{U_{ave}^2 + V_{ave}^2} \qquad (14)$$

Here, the wind speed and wind energy to z axis direction are ignored because the rotor of wind turbine cannot move toward z axis direction and wind energy to z axis direction cannot be utilized. In estimation of power generation, the wind energy at the point whose $U_{h,ave}$ is below 3 m/s is omitted, because the cut-in wind speed of AEOLOS wind turbine of 50 kW class is 3 m/s.

The power curve of AEOLOS wind turbine of 50 kW class is shown in Figure 4. The authors derive the empirical equation from the data of power curve, which is provided by AEOLOS. Figure 4 indicates the relationship between wind and power, resulting that the power generated by this wind turbine can be estimated by using the power curve. The power curve which is adopted in this study is as follows:

$$P_w = 59.075U_{h,ave}^3 - 62.619U_{h,ave}^2 - 33.433U_{h,ave} \quad (3\ m/s \le U_{h,ave} \le 10\ m/s) \qquad (15)$$

$$P_w = -793.94U_{h,ave} + 61.012\ (10\ m/s < U_{h,ave} \le 19\ m/s) \qquad (16)$$

where P_w is the power [in W] of wind turbine.

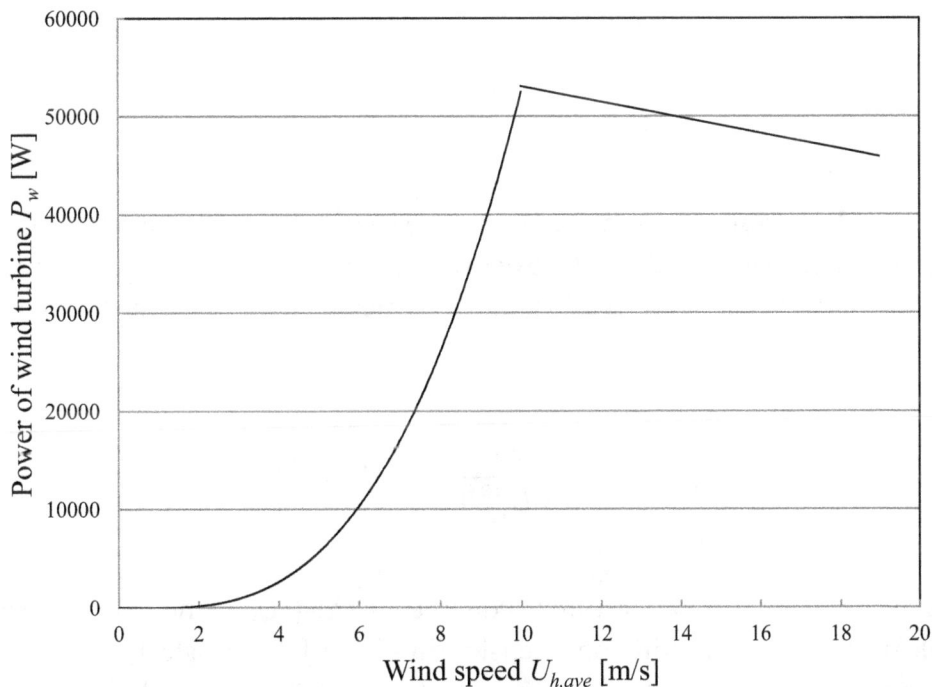

Figure 4. Power curve of wind turbine

2.4. Estimation of power generated from PV system

The power generated by PV system is calculated by using the following equation [39]:

$$E_p = H \times K \times P_p \div 1 \tag{17}$$

where E_p is the annual electric energy of PV [in kWh], H is the amount of solar radiation [in kWh/(m²)], K is the power generation loss factor, P_p is the system capacity of PV [in kW], 1 is the solar radiation intensity [in kW/m²] under standard state (AM1.5, solar radiation intensity: 1 kW/m², module temperature: 25 degree Celsius). In this study, the high performance PV HIT-B205J01 produced by Panasonic whose module conversion efficiency and maximum power per module are 17.4 % and 205 W, respectively is adopted for PV system [40]. The size of PV module is 1319 mm × 894 mm × 35 mm. P_p is calculated by installing this PV module on a roof of the building model, which is 67.7 kW$_p$. K is calculated by using the following equation:

$$K = K_p \times K_m \times K_i \tag{18}$$

where K_p is the power conversion efficiency of power conditioner, K_m is the correction factor decided by module temperature, and K_i is the power generation loss by interconnection and dirty of module surface. In this study, K_p and K_i are set at 0.945 and 0.95, respectively. K_p is assumed by referring to the performance of commercial power conditioning device SSI-TL55A2 manufactured by Panasonic [41]. K_m is calculated by the following equation:

$$K_m = 1 - \frac{(T_m - T_s)}{100} C \tag{19}$$

where T_m is the PV module temperature [in degree Celsius], T_s is the temperature [in degree Celsius] under standard test condition (=25 degree Celsius), and C is the temperature correction factor [in %/degree Celsius] which is 0.35 [42]. T_m is calculated by using the following equation [43]:

$$T_m = T_a + \left(\frac{46}{0.41U_m^{0.8} + 1} + 2 \right) I - 2 \tag{20}$$

where T_a is the ambient air temperature [in degree Celsius], U_m is the wind velocity [in m/s] over module of PV, I is solar radiation intensity [in kW/m²]. In this study, the meteorological data, such as solar radiation intensity, the ambient air temperature, and wind velocity of Tsu city in Japan are used from the data base METPV-11 provided by the New Energy and Industrial Technology Development Organization in Japan [44].

3. Results and discussion

3.1. Wind speed distribution around buildings in built environment

The contours of wind speed distribution in x direction (U) around the buildings for U_0=10 m/s at z=30 m are given in Figure 5 and they are on $x - y$ cross section of the building. It shows the distribution of U in case of β=0 degree, (i.e. the model faces the main wind direction). In this model, x=0 m and y=0 m is located at the center of distance between the nearest edge of adjacent two buildings. In Figure 5, the black lines mean the separation lines, which distinguish the different calculation domain in the model used for numerical simulation. It has been observed that the wind is accelerated within the intervening space between the buildings because some wind is over the U_0 of 10 m/s. The contracted flow occurs by passing through two buildings located like nozzle.

To investigate the location point of wind turbine, the contours of wind speed U distribution for U_0=10 m/s in the swept rotor area at the back of buildings for x/L=1.25, 1.875, and 2.50 on $y - z$ cross section are analyzed and it is presented in Figure 6. In the Figure 6, the black cross line shows the blades of wind turbine, if the wind turbine is located there. The black block lines show the building wake position. Although the wind speed decreases in the building wake, the wind is accelerated within the intervening space between the buildings at the area of the back of buildings for x/L=1.25, 1.875, and 2.50.

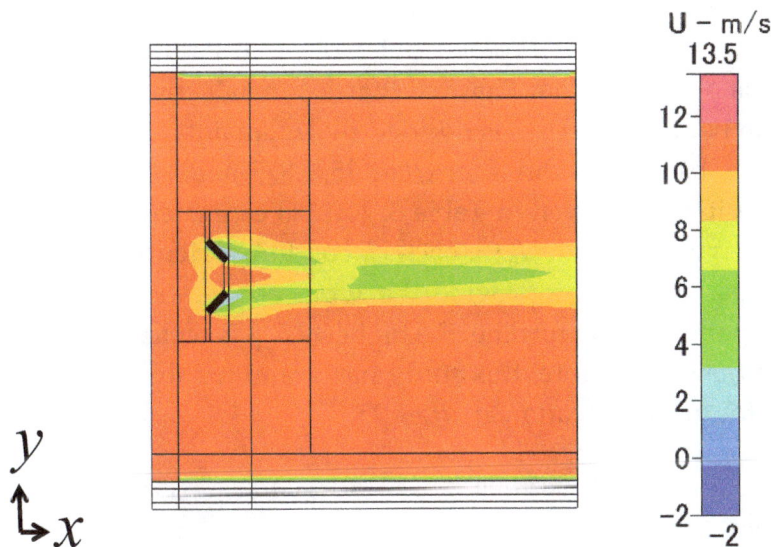

Figure 5. Contour of wind speed U distribution around buildings at z=30 m on $x - y$ cross section (U_0=10 m/s)

The wind speed distribution in the swept rotor area of wind turbine is important for estimating the wind turbine power output. The frequency distribution of U in the swept rotor area at x/L=1.25, 1.875, and 2.50 is given in Figure 7. It has been observed that the $U > U_0$ of 10 m/s is

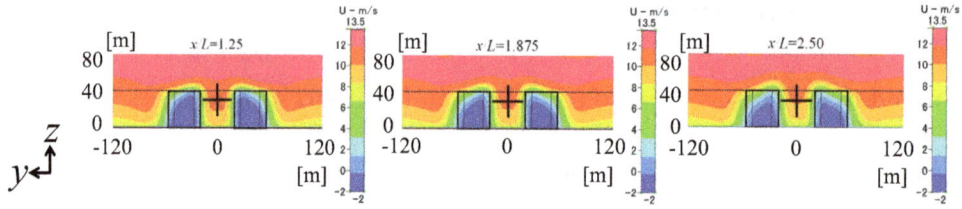

Figure 6. Contour of wind speed U distribution at the area at the back of buildings for x/L=1.25, 1.875, 2.50 on $y-z$ cross section (U_0=10 m/s)

confirmed at the area at the back of buildings of x/L=1.25, 1.875, and 2.50. Additionally, it is known that the higher U is obtained near the buildings.

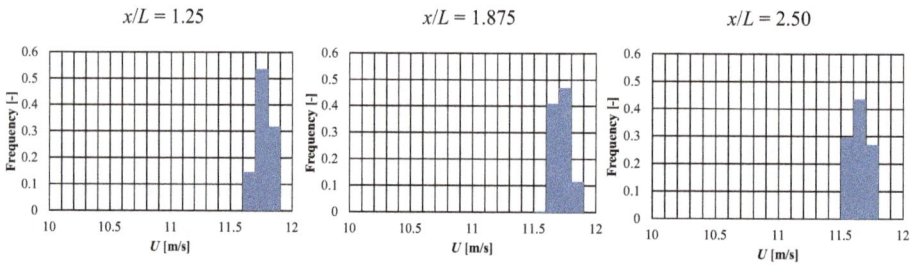

Figure 7. Frequency distribution of wind speed U in the swept rotor area at x/L=1.25, 1.875, and 2.50 (U_0=10 m/s)

This study carries out the 3D simulation and investigates the wind speed distribution toward y direction as well as x direction in order to calculate $U_{h,ave}$. The frequency distribution of wind speed towards y direction (V) in the swept rotor area for U_0=10 m/s at x/L=1.25, 1.875, and 2.50 are given in Figure 8. It is observed from Figure 8 that V is small compared to U shown in Figure 6. Therefore, it can be said that $U_{h,ave}$ is decided by U_{ave} mainly. The variations of $U_{h,ave}$ in the swept rotor area at the back of buildings for x/L=1.25, 1.875, and 2.50 in case of β=0 degree with the different U_0 condition are given in Table 3. $U_{h,ave}$ is estimated from the simulation results. It is seen that $U_{h,ave}$ is greater than U_0 for each U_0 condition. Hence, it can be concluded that the proposed building model can provide the wind acceleration irrespective of U_0. Considering the location point of wind turbine, the highest $U_{h,ave}$ is obtained at x/L=1.25 under these investigation conditions. Therefore, this study has examined the wind turbine power generation performance for turbine location at x/L=1.25.

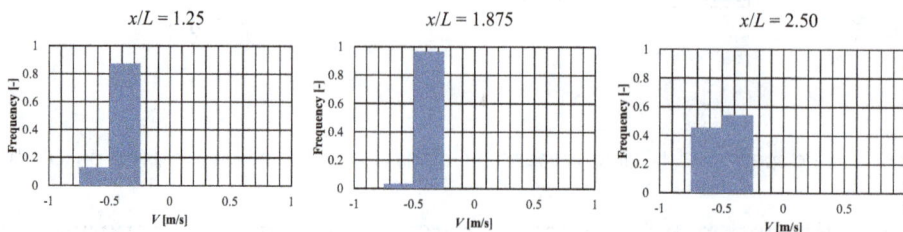

Figure 8. Frequency distribution of wind speed V in the swept rotor area at x/L=1.25, 1.875, and 2.50 (U_0=10 m/s)

U_0 [m/s]	3.00	4.00	5.00	6.00	7.00	8.00	9.00	10.00	11.00	12.00
$U_{h,ave}$ at x/L = 1.25 [m/s]	3.23	4.32	5.41	6.50	7.58	8.67	9.76	10.84	11.93	13.02
$U_{h,ave}$ at x/L = 1.875 [m/s]	3.15	4.22	5.28	6.35	7.41	8.47	9.54	10.60	11.67	12.73
$U_{h,ave}$ at x/L = 2.50 [m/s]	3.04	4.07	5.10	6.13	7.16	8.19	9.22	10.25	11.28	12.31

Table 3. $U_{h,ave}$ in case of β=0 degree under different U_0 conditions

3.2. Power generation performance of wind turbine located at proposed building layouts in actual area of Tsu city (Japan)

In order to decide the location of the wind turbine in proposed building layouts, the wind directions are considered (Figures 2 and 3) in the actual area of Tsu city. Annual wind direction distribution in Tsu city is given in Figure 9 and it is based on the wind speed data provided by the Japan Meteorological Agency [37]. The hourly measurement data from 2007 to 2011 are used for estimation of annual wind direction distribution.

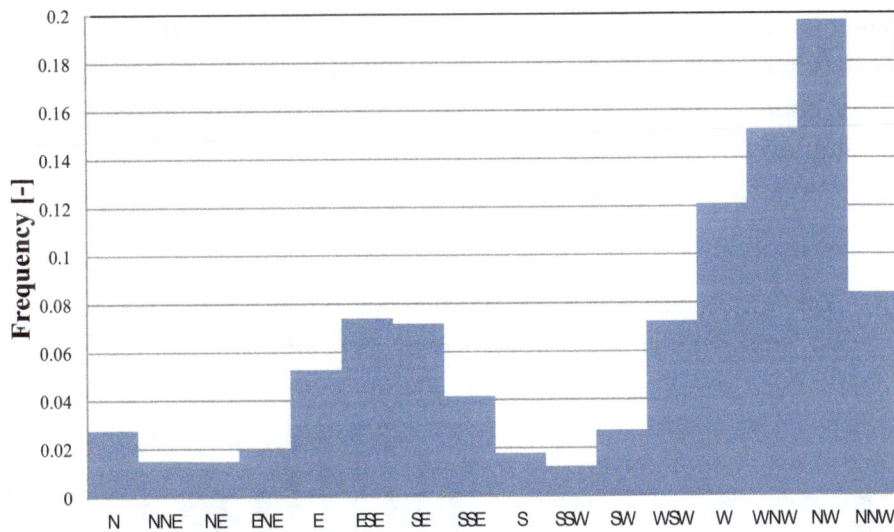

Figure 9. Annual wind direction distribution in Tsu city

It is observed from Figure 9 that the main wind direction throughout the year is North-West (NW). In this study, it is assumed that the open tip of building model is located as facing with the main wind direction (refer to Figure 3). West-North-West (WNW), North-North-West (NNW) and North (N) in addition to North-West (NW) are the wind directions, which can blow the wind among the buildings through nozzle. In this study, it is assumed that the wind blowing from the directions except for the above described five wind directions cannot be utilized for power generation of wind turbine. The different inflow angle conditions are considered in the simulation for finding the direction of wind in the proposed building layouts. As an example of the simulation results, the contours of wind speed U distribution around buildings for U_0=10 m/s at z=30 m on x-y cross section for angular inflow conditions are given in Figure 10. Although the wind blows through the intervening space between the buildings, the wind acceleration is not high. Hence, the wind power generation through wind speeds coming from the main wind direction is important.

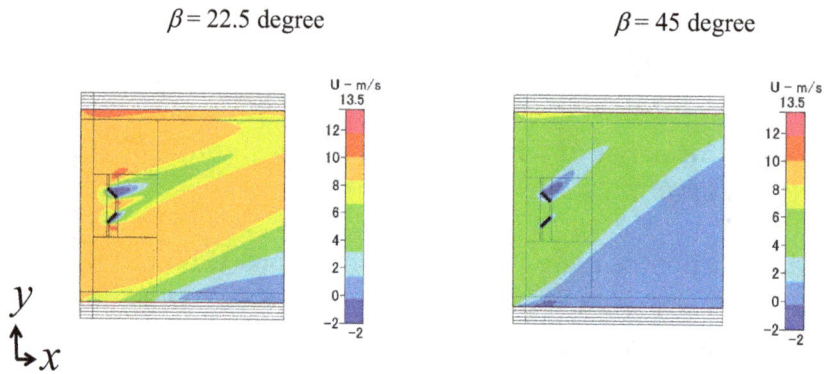

Figure 10. Contour of wind speed U distribution around buildings at $z=30$ m on $x - y$ cross section for angular inflow conditions (U_0=10 m/s)

The hourly data on wind speed and direction are used as inputs in the simulation for finding the wind turbine power output at a location of wind turbine in the proposed building layouts. The daily power energy outputs of wind turbine, which is installed at the location of the proposed building layout, for months January, April, July, and October are given in Figures 11, 12, 13, and 14, respectively. These four months are considered as representative of four seasons.

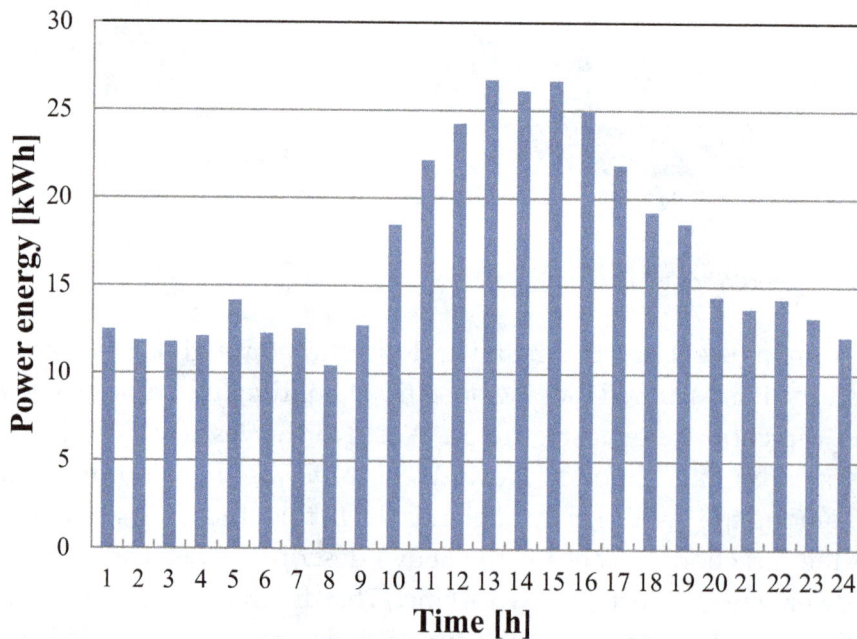

Figure 11. Variation of wind energy output in January in case of installing proposed building layouts in Tsu city (x/L=1.25)

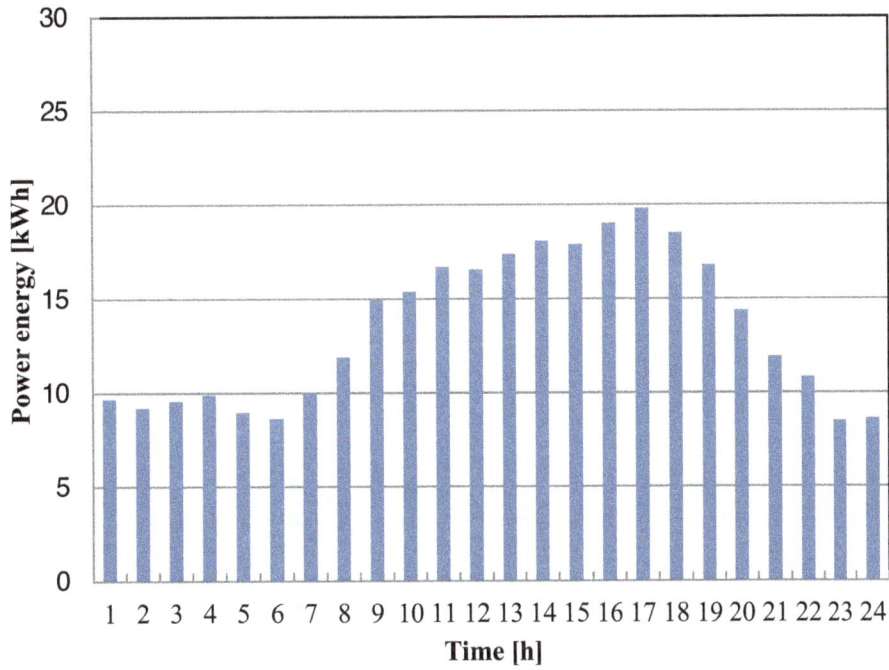

Figure 12. Variation of wind energy output in April in case of installing proposed building layouts in Tsu city (*x*/*L*=1.25)

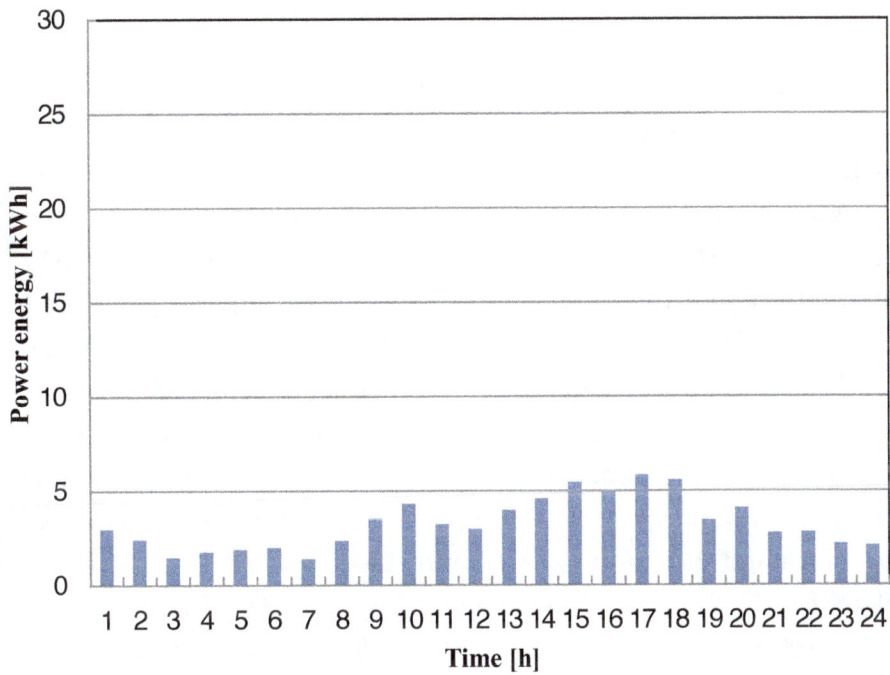

Figure 13. Variation of wind energy output in July in case of installing proposed building layouts in Tsu city (*x*/*L*=1.25)

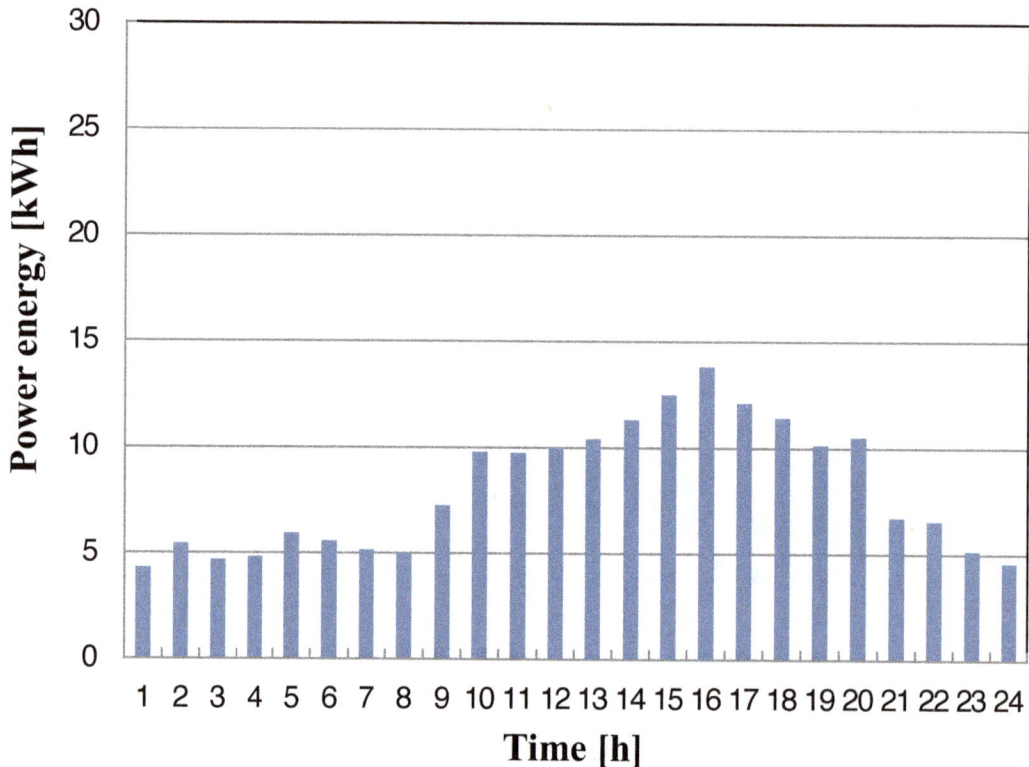

Figure 14. Variation of wind energy output in October in case of installing proposed building layouts in Tsu city (*x/L*=1.25)

From these figures, it is observed that higher power energy of wind turbine can be obtained in the daytime irrespective of month/season. In addition, the higher amount of total wind energy through a day is obtained in January, while the lower amount of total wind energy through a day is obtained in July (in comparison with these four representative months). The main wind direction in July is East-South-East (ESE) and it is shown in Figure 15, while the main wind direction throughout the year is North-West (NW). In this study, it is assumed that the winds blowing from the restricted wind directions can be utilized for power generation from wind turbine. The restricted wind directions mean five directions whose center is the main wind direction, and the other four directions are located symmetry to the main wind direction. The amount of wind energy production is estimated to be zero for the wind blowing from the other directions. The monthly main wind direction is East-South-East (ESE) in July as shown in Figure 15, which is different from the annual main wind direction. The frequency of wind blowing from the restricted five directions is small in July, resulting that the wind energy production is lower compared to the other months. Therefore, while selecting the building orientations, it is very important to consider the wind speed directions in order to maximize annual wind energy production.

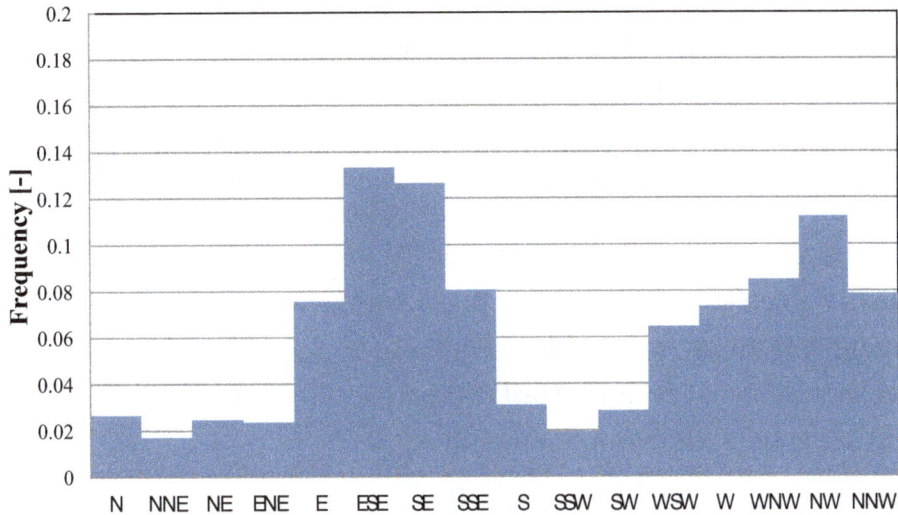

Figure 15. Frequency distribution of wind direction in July in Tsu city

3.3. Power generation performance of PV system located at proposed building layouts in actual area of Tsu city (Japan)

To maximize the power output of PV system, it is important to set the tilt angle of solar panel normal to the solar orbit. Hence, this study investigates the optimum tilt angle for installing PV system on roof of the proposed building model in Tsu city. By examining the data of solar radiation intensity in Tsu city for five year from 1999 to 2009 [44], the best tilt angle is 35 degree (Figure 16). Therefore, this study estimates the power output of PV system installed on roof of a building whose tilt angle is 35 degree. In addition, this study assumes that the PV arrays face to the south.

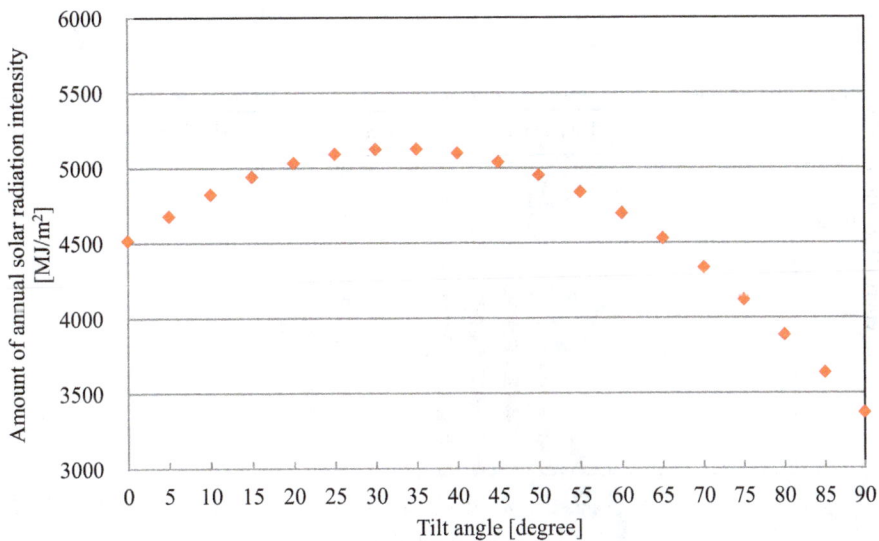

Figure 16. Effect of tilt angle of PV array assumed to be installed in Tus city on amount of annual solar radiation intensity

The daily power outputs from PV system, for months January, April, July, and October, are given in Figures 17, 18, 19, and 20, respectively. The hourly meteorological measurement data of solar radiation intensity from 1990 to 2009 [44] and that of temperature and wind velocity from 2010 to 2011 [37] which are averaged among all days in each month are used for estimation of daily power outputs of PV system.

Figures 17, 18, 19, and 20 provide the output of PV system with respect to the time. Though the solar radiation intensity in July is high, the power generation from PV system is reduced due to high temperature (as explained by Eq. (19)). Because the solar radiation is relatively high and temperature is not high in April compared to the other months/seasons, therefore the PV system has good performance in month April.

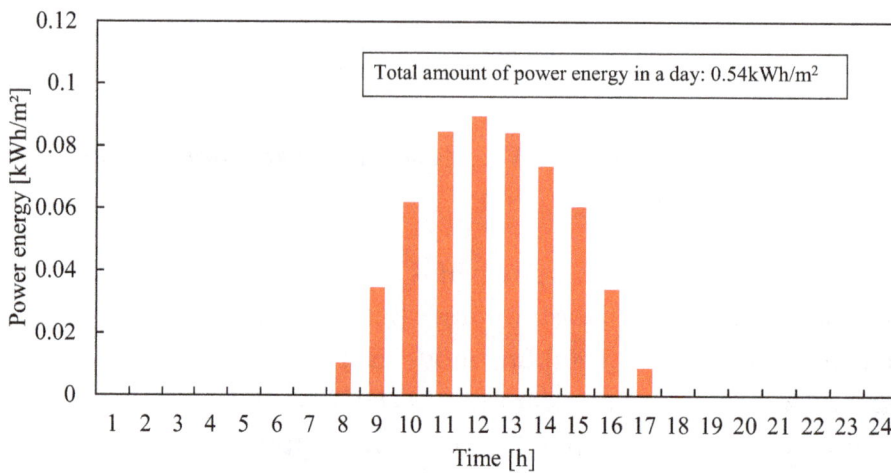

Figure 17. Variation of PV energy output in January in case of installing in Tsu city

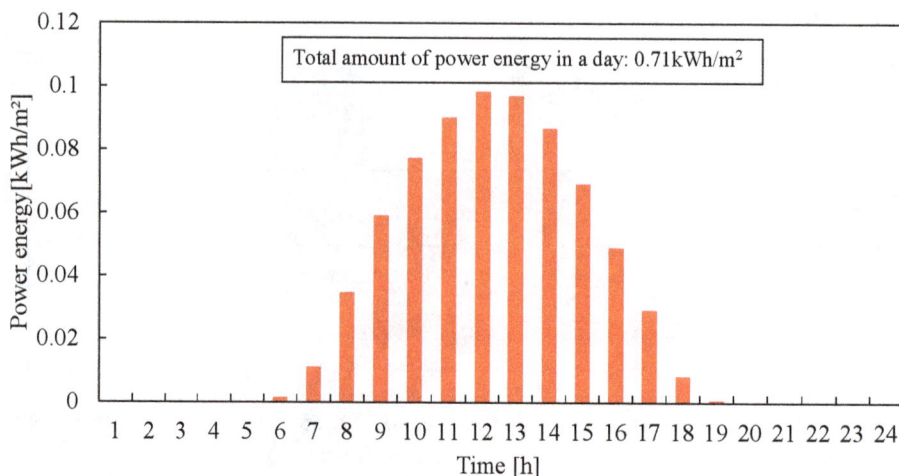

Figure 18. Variation of PV energy output in April in case of installing in Tsu city

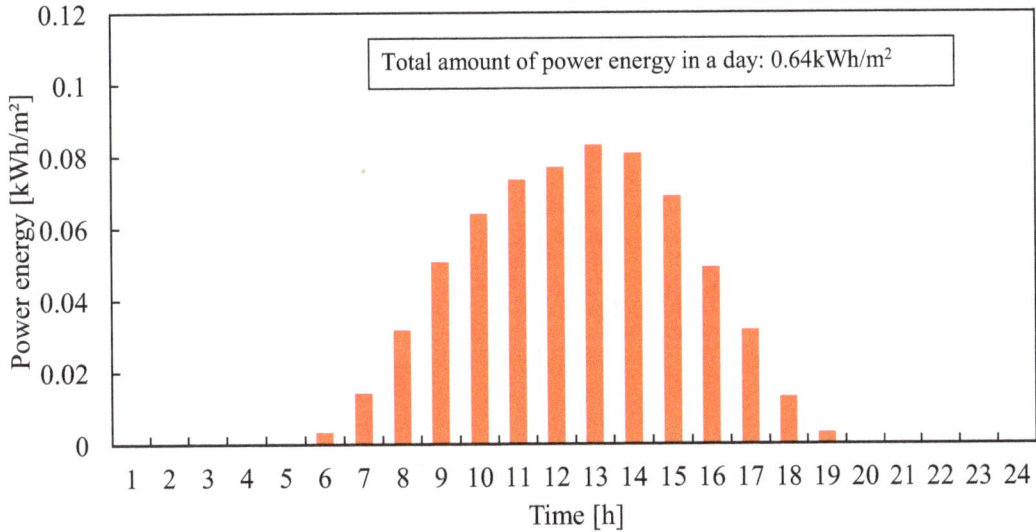

Figure 19. Variation of PV energy output in July in case of installing in Tsu city

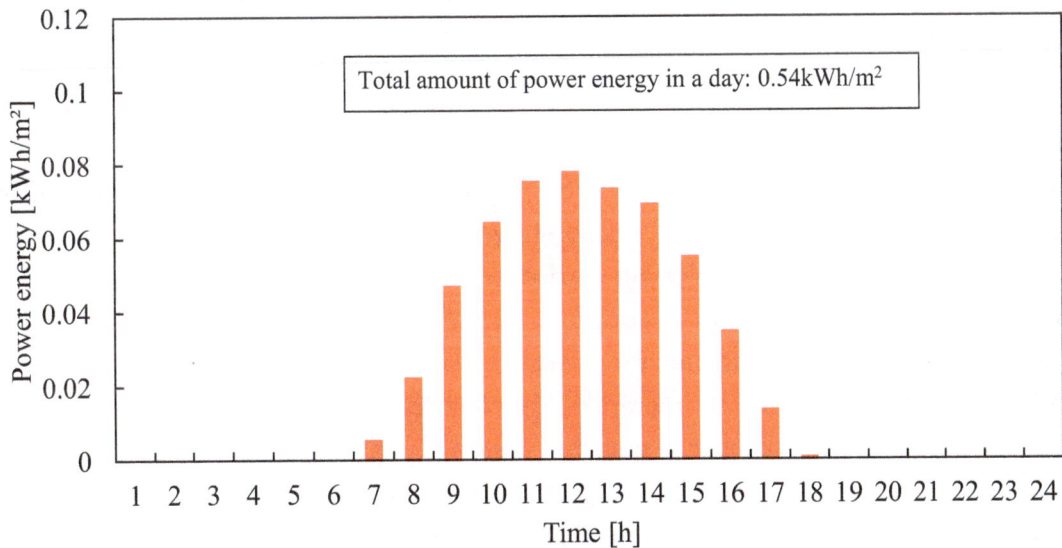

Figure 20. Variation of PV energy output in October in case of installing in Tsu city

3.4. Investigation on performance of renewable energy supply system in built environment

This study investigates the power generation performance of a wind turbine and PV system integrated with two buildings and the power supply characteristics from this combined power generation system with electric demand are presented. In the building model of this study, one wind turbine and two PV arrays of 67.7 kWp per two buildings are assumed to be installed.

As described above, by using the meteorological data [37, 44], the daily power outputs of wind turbine and PV system for each month throughout a year and annual power outputs from them are estimated. As a result, the annual electrical energy production of combined power generation system is 153 MWh. A typical monthly electric consumption for a household in Japan (for year 2012) is 276 kWh [45], resulting that the annual electric consumption for the proposed building model, which has 80 households per two buildings, is 265 MWh. Therefore, this combined power generation system can cover the 57.7% of the annual electric consumption of households assumed to be living in the building model. In the future study, the change of self-sufficient ratio with time will be investigated by collecting the statistical data of electric consumption.

To compare the power supply characteristics of this combined power generation system with the electric demand characteristics by the other consumer, the energy data base of Mie university [46] which is located in Tsu city is adopted. The data of daily electric consumption in weekday from 2010 to 2011 is used from this data base. Then, the daily data of a representative one day in each month is estimated by averaging all daily data through week days in each month. In this estimation, the electric consumption of Mie university per floor space is calculated from the data base and applied to the floor space of the building model, resulting that the amount of daily electric consumption for this building model is derived.

Figures 21, 22, 23, and 24 show the daily power outputs of wind turbine, PV system, and combined them for months January, April, July, and October. Additionally, the daily self-sufficient ratio of combined power generation system, which is a ratio of electrical energy generation to electric consumption, is also shown. From these figures, it is clear that the electrical energy of combined power generation system as well as that of wind turbine or PV system in daytime is larger than nighttime. As the electric consumption of university in daytime is also larger than nighttime, the characteristics of power supply by combined power generation system matches the characteristics of electric consumption of university. The self-sufficient ratio of combined power generation system in daytime is approximately 20 – 60 %, while that in nighttime is below 20 %. The average self-sufficient ratios of combined power generation system in all the day for January, April, July, and October are 23.3 %, 30.0 %, 15.7 %, and 19.9 %, respectively. The highest self-sufficient ratio of combined power generation system is obtained in April. Because April is the moderate season in Japan, the electric consumption is relatively small compared to summer and winter. In addition, the amount of power energy of PV system in April is higher compared to the other months/seasons as described above. Consequently, the self-sufficient ratio of combined power generation system in April is the best. On the other hand, the self-sufficient ratio of combined power generation system in July is lower compared to the other months/seasons. In July, the electric consumption is high due to hot season in Japan. In addition, the power energy of combined power generation system is low because the meteorological condition is not good for wind turbine and PV system as described above. Therefore, the self-sufficient ratio of combined power generation system in July shows the small value relatively.

Self-sufficient ratio [%]

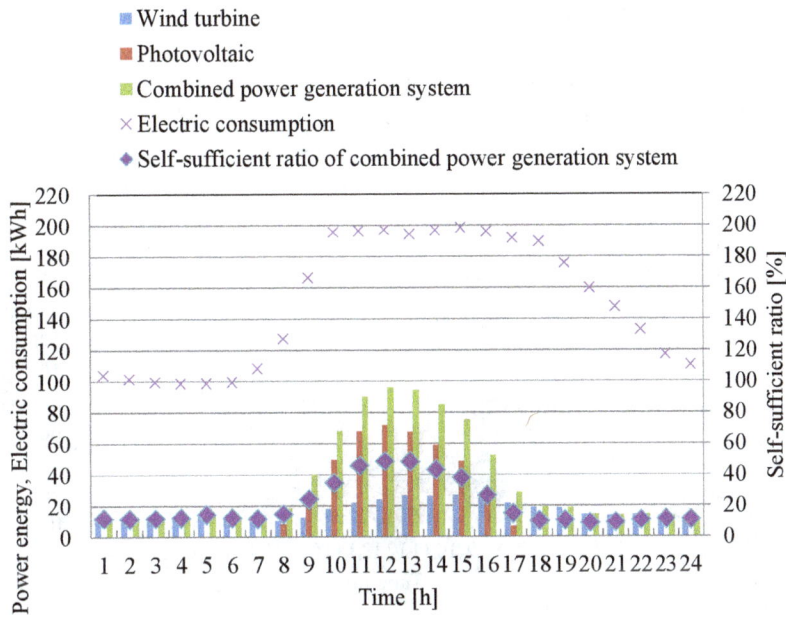

Figure 21. Variation of energy output of each power generation system and self-sufficient ratio of power generation to electric consumption in January

Self-sufficient ratio [%]

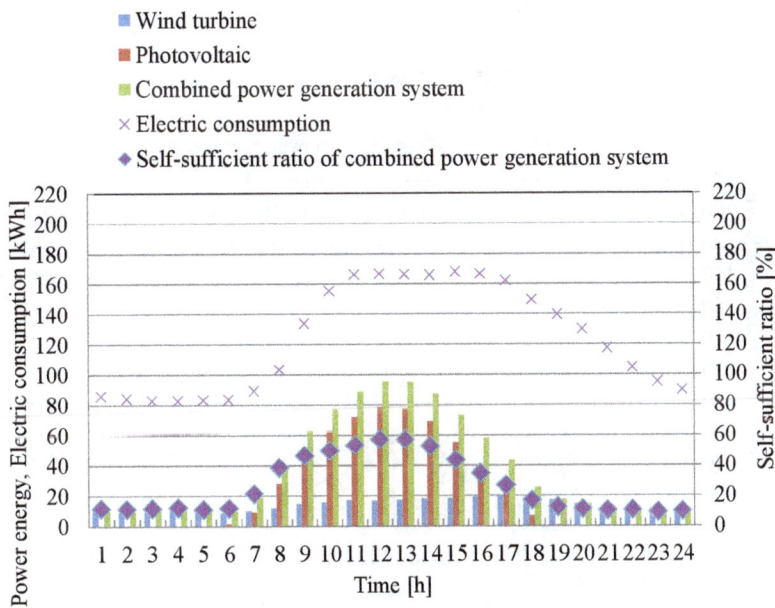

Figure 22. Variation of energy output of each power generation system and self-sufficient ratio of power generation to electric consumption in April

Self-sufficient ratio [%]

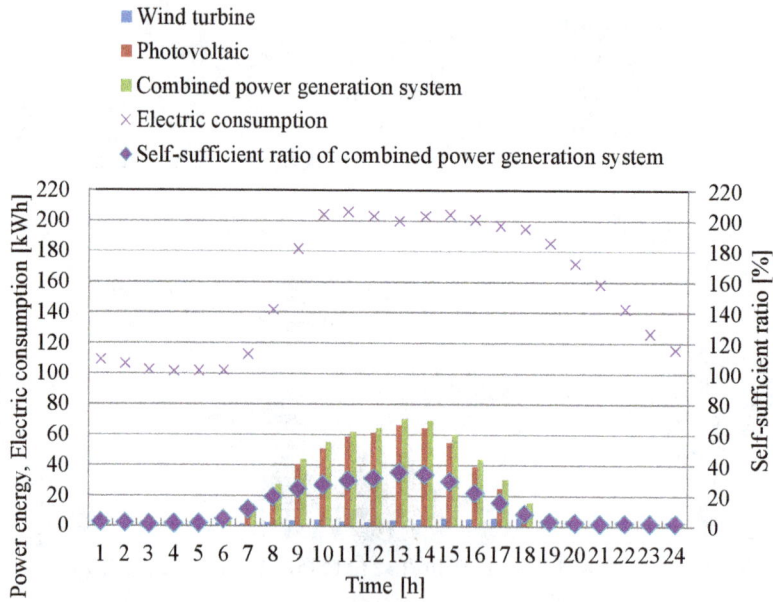

Figure 23. Variation of energy output of each power generation system and self-sufficient ratio of power generation to electric consumption in July

Self-sufficient ratio [%]

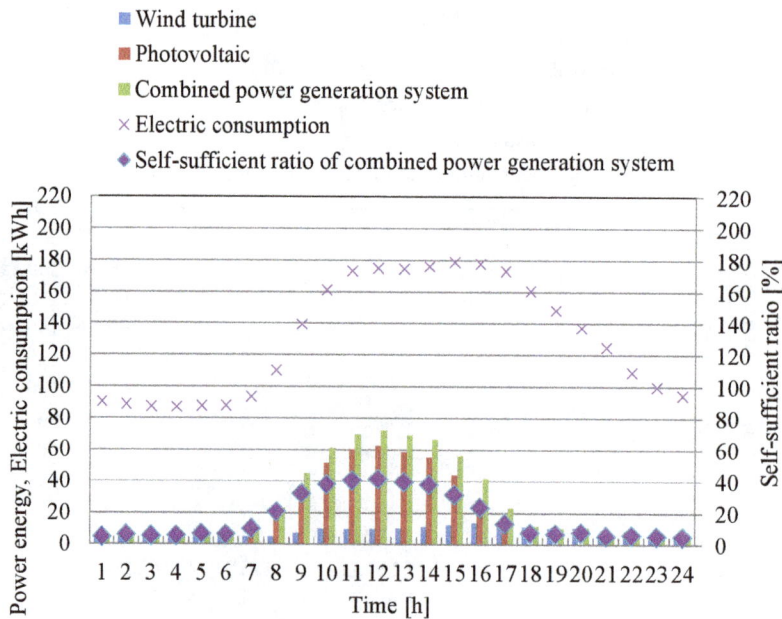

Figure 24. Variation of energy output of each power generation system and self-sufficient ratio of power generation to electric consumption in October

The proposed model of combined built environment renewable energy supply system is effective for reducing the usage of the power generated by thermal power plant, i.e., it is effectively reducing the significant amount of CO_2 emission. To improve the performance of the proposed building model, the following investigations are suggested:

i.   The effect of building size and layout on the wind speed distribution around buildings and power generation performance of built environment wind turbine should be investigated. For example, though the angle between two buildings of 90 degree is proved to be effective for accelerating the wind, the other angles have a possibility to produce higher power from wind turbine. As an example, the wider angle such as 135 degree can suck the wind inside the nozzle shaped building model. If the speed of accelerated wind obtained by the building model whose angle between two building is 135 degree is comparable to that by the building model whose angle between two building is 90 degree, the power generated by wind turbine is larger due to increase in the number of wind direction whose wind can be utilized for power generation. However, it is believed that there is the optimum building size and layout to obtain higher wind acceleration effectively because it is difficult to make a contracted flow by too wider angle between two buildings.

ii.   To select the area where the wind direction is not changed throughout the year is necessary to obtain the high accelerated wind by the proposed building model. The wind blowing from the main wind direction is the most desirable to realize the high performance of the proposed building model. Therefore, the research on the optimum area for installing the building model is important. Otherwise, the building layout should be changed by increasing the number of buildings. If the four buildings are located like a cross with keeping the space at the center for installing a wind turbine, the wind from every wind direction might be tapered.

iii.   Though the tilt angle of PV array is optimized by examining the data of solar radiation, the installation direction of PV array is fixed opening towards south. If the system tracking solar orbit is considered in the proposed building model, the power output of PV system increases. In addition, the effect of installing PV modules on side wall of buildings is effective for increasing the power generated from PV system.

iv.   Feasibility study on the proposed building model by assuming the installation of it in more different actual areas should be carried out. By investigating the versatility of the proposed building model under the various meteorological conditions, the assumption and building size and layout which are set in the model would be improved to match the actual meteorological condition.

v.   As an application model to meet the electric demand more, the building model including secondary battery and fuel cell systems as well as wind turbine and PV system can be imaginable. If there is a miss match between the power supply of combined renewable power generation system such as wind turbine and PV and electric consumption, the secondary battery or fuel cell system can provide the power to cover the miss match. In order to produce the hydrogen for fuel cell, the solar thermal chemical conversion from ammonia to hydrogen is also promising. The renewable energy can be stored as an electro-chemical energy and it can be used as energy source when it is needed. Solar thermal chemical conversion can also help in overcoming the miss match between the power supply of combined renewable power generation systems and electric demand.

4. Conclusions

In this chapter, building topologies/orientations/layouts in a smart city are investigated for finding out the wind speed distribution profiles in the built environment. The analysis of wind speed distribution and directions are very important for not only to find the mechanical wind stress but also to find the energy content in the wind and a location for wind turbine in built environment. This analysis is also useful for designing the building layouts in such a way to make the nozzle of the wind by using wind directions and then finding out the proper location of the wind turbine in smart city. In this work, building layouts like nozzle is proposed and investigated to obtain the contracted flow by blowing wind through the buildings. The output power of wind turbine is estimated by using the power curve of real wind turbine and the wind speed distribution around buildings by using the wind speed data for Tsu city (Japan). This chapter also investigates the power generation performance of PV system installed on a roof of the building by using the actual meteorological data of Tsu city. The power generation characteristics of the combined system including wind turbine and PV assumed to be operated under the actual meteorological conditions, is evaluated and compared with the electric consumption profile of the consumer assumed to utilize/occupy the building. The main conclusions obtained from these investigations are as follows:

i.   In case of β=0 degree, the wind is accelerated within the intervening space between the buildings in the back area of buildings for x/L=1.25, 1.875, and 2.50, and it is observed that $U > U_0$ (of 10 m/s) is established in the proposed area.

ii.   Because $U_{h,ave}$ at the back area of buildings for x/L=1.25, 1.875, and 2.50 is higher than U_0 in case of β=0 degree, and hence the proposed building layouts can provide the wind speed acceleration irrespective of U_0.

iii.   In case of installing the building model in Tsu city, the higher wind energy output is obtained in the daytime irrespective of the month/season. The higher wind energy output throughout the day is available in January, while the lower wind energy output is available in July.

iv.   In case of installing the building model in Tsu city, the highest total amount of power energy generated by PV system in throughout the day is obtained in month April as the solar radiation is relatively high and temperature is not high compared with the other months/seasons.

v.   The combined power generation system proposed by this study can cover the 57.7% of the annual electric consumption of households assumed to be living in the building model.

vi.   Comparing the power generation characteristics of the combined power generation system with the electric consumption characteristics of university, it is known that the self-sufficient ratio of combined power generation system in daytime is approximately 20 – 60 %, while that in nighttime is below 20 %. The highest self-sufficient ratio of combined power generation system is available in April, while the lowest self-sufficient ratio is available in July.

vii. In order to realize the energy supply system with CO_2 free by improving the power generation performance of the proposed building model and energy supply adaptability to energy demand, the further investigation such as the optimization of building configuration/layouts and modeling of new combined system including power generation, energy preservation, and energy conversion is needed in near future.

Author details

Akira Nishimura[1*] and Mohan Kolhe[2]

*Address all correspondence to: nisimura@mach.mie-u.ac.jp

1 Division of Mechanical Engineering, Graduate School of Engineering, Mie University, Tsu, Japan

2 Faculty of Engineering & Science, University of Agder, Grimstad, Norway

References

[1] Ministry of Economy, Trade and Industry in Japan. http://www.meti.go.jp/english/policy/energy_environment/smart_community/index.html (accessed 3 June 2014).

[2] Smart Grid Demonstration Project in Sino-Singapore Tianjin Eco-City. http://www.sgiclearinghouse.org/Asia?q=node/2594&lb=1 (accessed 3 June 2014).

[3] Kolhe M. Smart Grid: Charting a New Energy Future: Research, Development and Demonstration. The Electricity Journal 2012; 25: 88-93.

[4] Yoshie R, Tominaga Y, Mochida A, Kataoka H, Yoshikawa Y. CFD Simulation of Flow-field around a High-rise Building Located in Surrounding City Blocks, Part (1)

Outline of Experiment and Calculation, Influence of Various Calculation Conditions. Wind Engineers 2005; 30(2): 133-134.

[5] Murakami S, Hibi K, Mochida A. Three Dimensional Analysis of Turbulent Flowfield around Street Blocks by Means of Large Eddy Simulation (Part II), (Investigation of the Relation between Fluctuating Pressure and Flowfield on and around Building). J. Archit. Plann. Environ. Engng. 1991; 425: 11-19.

[6] Kataoka H, Mizuno M. Numerical Simulation of Separating Flow around a Body Using Artificial Compressibility Method. J. Archit. Plann. Environ. Eng. 1998; 504: 63-70.

[7] Nishimura K, Yasuda R, Ito S. An Experimental and Numerical Study of Concentration Prediction around a Building, Part II Numerical Simulation by k-ε Model. J. Jpn. Soc. Atmos. Environ. 1999; 34(2): 103-122.

[8] Nozu T, Tamura T. Application of Computational Fluid Technique with High Accuracy and Conservation Property to the Wind Resistant Problems of Buildings and Structures, Part 3 Analysis of the Flows and the Wind Pressure around a High-rise Building in the Turbulent Boundary Layer. J. Struct. Constr. Eng. 2000; 538: 65-72.

[9] Nozu T, Tamura T. Application of Computational Fluid Technique with High Accuracy and Conservation Property to the Wind Resistant Problems of Buildings and Structures, Part 2 Vortex Structures and Aerodynamic Characteristics around a Three-dimensional Square Prism on a Ground Plane. J. Struct. Constr. Eng. 1998; 503: 37-43.

[10] Kose D A, Fauconnier D, Dick E. ILES of Flow over Low-rise Buildings: Influence of Inflow Conditions on the Quality of the Mean Pressure Distribution Prediction. J. Wind. Eng. Ind. Aerodyn, 2011; 99: 1056-1068.

[11] Meng Y, Hibi K. Turbulence Characteristics and Organized Motions on the Flat Roof of a High-rise Building. J. Wind Engineering 1997; 72: 21-34.

[12] Meng Y, Oikawa S. A Wind-tunnel Study of the Flow and Diffusion within Model Urban Canopies, Part 1. Flow Measurements. J. Jpn. Atmos Environ. 1997; 32(2): 136-147.

[13] Oikawa S, Ishihara T, Yasuda R, Nishimura K, Hase M. An Experimental and Numerical Study of Concentration Prediction around a Building, Part I Wind-tunnel Study. J. Jpn. Soc. Atmos. Environ. 1999; 34(2): 123-136.

[14] Zaki S A, Hagishima A, Tanimoto J. Experimental Study of Wind-induced Ventilation in Urban Building of Cube Arrays with Various Layouts. J. Wind. Eng. Ind. Aerodyn. 2012; 103: 31-40.

[15] Ouammi A, Dagdougui H, Sacile R, Mimet A. Monthly and Seasonal Assessment of Wind Energy Characteristics at Four Monitored Locations in Liguria Region (Italy). Renewable and Sustainable Energy Reviews 2010; 14: 1959-1968.

[16] Oyedepo S O, Adaramola M S, Paul S S. Analysis of Wind Speed Data and Wind Energy Potential in Three Selected Locations in South-east Nigeria. Int. J. Ene. Environ. Eng. 2012; doi: 10.1186/2251-6832-3-7.

[17] Olaofe Z O, Folly K A. Wind Energy Analysis Based on Turbine and Developed Site Power Curves: A Case-study of Darling City. Renewable Energy 2013; 53: 306-318.

[18] Saito S, Sato K, Sekizuka S. A Discussion on Prediction of Wind Conditions and Power Generation with the Weibull Distribution. JSME Int. J. Series B 2006; 49(2): 458-464.

[19] Khadem S K, Hussain M. A Pre-feasibility Study of Wind Resources in Kutubdia Island, Bangladesh. Renewable Energy 2006; 31: 2329-2341.

[20] Firtin E, Guler O, Akdag S A. Investigation of Wind Shear Coefficients and Their Effect on Electrical Energy Generation. Applied Energy 2011; 88: 4097-4105.

[21] Rocha P A C, Sousa R C, Andrade C F, Silva M E V. Comparison of Seven Numerical Methods for Determining Weibull Parameters for Wind Energy Generation in the Northeast Region of Brazil. Applied Energy 2012; 89: 395-400.

[22] Usta I, Kanter Y M. Analysis of Some Flexible Families of Distributions for Estimation of Wind Speed Distributions. Applied Energy 2012; 89: 355-367.

[23] Mostafaeipour A. Feasibility Study of Offshore Wind Turbine Installation in Iran Compared with the World. Renewable and Sustainable Energy Reviews 2010; 14: 1722-1743.

[24] Wang L, Yeh T H, Lee W J, Chen Z. Analysis of a Commercial Wind Farm in Taiwan, Part I: Measurement Results and Simulations. IEEE Transactions on Industry Application 2011; 47(2): 939-953.

[25] Nishimura A, Ito T, Murata J, Ando T, Kamada Y, Hirota M, Holhe M. Wind Turbine Power Output Assessment in Built Environment 2013; 4(1): 1-10. doi: 10.4236/sgre. 2013.41001.

[26] Nishimura A, Ito T, Kakita M, Murata J, Ando T, Kamada Y, Hirota M, Holhe M. Impact of Building Layouts on Wind Turbine Power Output in the Built Environment: A Case Study of Tsu City 2014; 94: 315-322.

[27] Japan Meteorological Agency, http://www.data.jma.go.jp/obd/stats/etrn/indcx.php (accessed 17 June 2014).

[28] Wen C Y, Yang A S, Tseng L Y, Tsai W T. Flow Analysis of a Ribbed Helix Lip Seal with Consideration of Fluid-Structure Interaction. Computers & Fluids 2011; 40(1): 324-332. doi: 10.1016/j.compfluid.2010.10.005.

[29] Caicedo H H, Hernandez M, Fall C P, Eddington D T. Multiphysics Simulation of a Microfluidic Perfusion Chamber for Brain Slice Physiology. Biomed Micodevices 2010; 12(5): 761-767. doi: 10.1007/s10544-010-9430-5.

[30] Xing C, Braun M J, Li H. Damping and Added Mass Coefficients for a Squeeze Film Damper Using the Full 3-D Navier Stokes Equation. Tribology International 2010; 43(3): 654-666.

[31] Demirkaya G, Soh C W, Ilegbusi O J. Direct Simulation of Navier-Stokes Equations by Radial Basis Functions. Applied Mathematical Modeling 2008; 32(9): 1848-1858.

[32] Glatzel T, Litterst C, Cupeli C, Lindramann T, Moosmann C, Niekrawietz R, Sterule W, Zengerle R, Koltay P. Computational Fluid Dynamics (CFD) Software Tools for Microfluidic Applications – A Case Study. Computers & Fluids 2008; 37(3): 218-235.

[33] Kabir M A, Khan M M K, Rasul M G. Flow of a Mixed Solution in a Channel with Obstruction at the Entry: Experimental and Numerical Investigation and Comparison with Other Fluids. Experimental Thermal and Fluid Science 2006; 30(6): 497-512.

[34] ESI, editor. CFD-ACE+Modules Manual Part 1. Huntsville: ESI CFD Inc.; 2009.

[35] ESI, editor. CFD-ACE+Modules Manual Part 2. Huntsville: ESI CFD Inc.; 2009.

[36] AEOLOS wind turbine. http://www.windturbinestar.com/ (accessed 4 June 2014).

[37] Japan Meteorological Agency. http://www.data.jma.go.jp/obd/stats/etrn/index.php (accessed 4 June 2014).

[38] Ministry of Internal Affairs and Communications in Japan. http://www.e-stat.go.jp/SG1/estat/ListE.do?bid=000001029530&cycode=0 (accessed 4 June 2014).

[39] New Energy and Industrial Technology Development Organization in Japan. http://www.nedo.go.jp/content/100110086.pdf (accessed 4 June 2014).

[40] Panasonic. http://sumai.panasonic.jp/solar/lineup.html (accessed 1 Jan 2013).

[41] Panasonic. http://panasonic.biz/energy/solar/index.html (accessed 1 Jan 2013).

[42] Kawamoto K, Nakatani S, Hagihara R, Nakai T, Baba T. High Efficiency HIT Solar Cell. Sanyo Technical Review 2002; 34(1): 111-117.

[43] Oozeki T, Izawa T, Otani K, Tsuzuku K, Koike H, Kurokawa K. An Evaluation Method for PV Systems by Using Limited Data Item. IEEJ Transactions on Power and Energy 2005; 125(12): 1299-1307.

[44] New Energy and Industrial Technology Development Organization in Japan. http://app7.infoc.nedo.go.jp/ (accessed 6 June 2014).

[45] Digital Pamphlet of Drawings on Atomic Power and Energy, http://fepc-dp.jp/?type=pub&ID=7&Chapter=1&page=32&spg=8&epg=32 (accessed 11 June 2014).

[46] The Energy Consumption Data Base of Mie University, Mie Taro, https://www.mietaro.com/mietaro/index.php (accessed 11 June 2014).

5

A Study of Various Aspects of Cement Chemistry and Industry Relevant to Global Warming and the Low Carbon and Low Energy Molten Salt Synthesis of Cement Compounds

Georgios M. Photiadis

1. Introduction

Global warming caused by the 'greenhouse effect' is mainly due to the CO_2 emissions from human activities such as fossil fuel use (3/4ers) and land use change (1/4er). The concentration of atmospheric CO_2 has increased from a pre-industrial value of ~280 ppm in 1750 to 391 ppm in 2011. The industrial production of Portland cement clinker involves mixing and heating the raw materials limestone and clay minerals in a rotary kiln up to 1450°C in a complex solid/liquid state reaction process. The decomposition of limestone, the combustion of fuels in the kiln and the power generation required for grinding the raw materials and the product, result in process and energy-related emissions of ~0.8 kg CO_2 / kg of cement produced. Thus, the cement industry contributes ~5% in the global anthropogenic CO_2 emissions. 'ULECeS' EPSRC-funded project involved research on the molten salt synthesis of cement, aiming to reduce significantly these CO_2 emissions.

1.1. Global warming, climate change and the intergovernmental panel on climate change

During the 19th century scientists believed that gases in the atmosphere of the earth cause a "greenhouse effect", thus having a direct effect on the temperature of the planet. A lower level of carbon dioxide in the atmosphere in the distant past was also linked to the ice ages of these periods [1].

At around 1900 the Swedish chemist Svante Arrhenius calculated that CO_2 emissions from the burning of fossil fuels and other combustion processes might someday bring a global warming

[2, 3]. Other scientists dismissed his idea as faulty. In the last few decades accumulating evidence points that the suggestion of Svante Arrhenius might not be faulty at all and that we may indeed face already the first consequences of global warming and of climate change. The need to take some kind of action to tackle global warming and climate change led to the foundation of the Intergovernmental Panel on Climate Change (IPCC) in order to start the study of these phenomena on a worldwide basis.

The tasks and the work undertaken by the IPCC might prove to be of enormous importance for the survival and the future of human species on planet Earth. The rather reluctant response of leading global governments to the suggestions of IPCC seems to be accompanied by an attitude of trying to transfer the responsibility for hard but urgent decisions to the next generations. It is of paramount importance the citizens around the globe to become aware of the work and the scope of IPCC. This is the main reason that in the following paragraphs some relevant information about IPCC will be given, bearing in mind that well informed citizens is the necessary condition to tackle phenomena of huge complexity such as global warming and climate change.

The Intergovernmental Panel on Climate Change (IPCC) is the leading international body for the assessment of climate change [4]. In the official internet site of the IPCC there is the following statement about its founding bodies as well as about its role [5]:

"It (the IPCC) was established by the United Nations Environment Programme (UNEP) and the World Meteorological Organization (WMO) in 1988 to provide the world with a clear scientific view on the current state of knowledge in climate change and its potential environmental and socio-economic impacts. In the same year, the UN General Assembly endorsed the action by WMO and UNEP in jointly establishing the IPCC. The IPCC is a scientific body under the auspices of the United Nations (UN). It reviews and assesses the most recent scientific, technical and socio-economic information produced worldwide relevant to the understanding of climate change. It does not conduct any research nor does it monitor climate related data or parameters. Thousands of scientists from all over the world contribute to the work of the IPCC on a voluntary basis. Review is an essential part of the IPCC process, to ensure an objective and complete assessment of current information. IPCC aims to reflect a range of views and expertise."

The aim of the IPCC to provide an authoritative international statement of scientific understanding of climate change is implemented by periodic assessments of the causes, impacts and possible response strategies to climate change. These are considered to be the most comprehensive and up-to-date reports available on the subject, and form the standard reference for all concerned with climate change in academia, government and industry worldwide [6].

The IPCC work is shared among three Working Groups, a Task Force and a Task Group [7]:

The IPCC Working Group I (WG I) assesses the physical scientific aspects of the climate system and climate change including: changes in greenhouse gases and aerosols in the atmosphere; observed changes in air, land and ocean temperatures, rainfall, glaciers and ice sheets, oceans and sea level; historical and paleoclimatic perspective on climate change; biogeochemistry, carbon cycle, gases and aerosols; satellite data and other data; climate models; climate projections, causes and attribution of climate change [7].

The IPCC Working Group II (WG II) assesses the vulnerability of socio-economic and natural systems to climate change, negative and positive consequences of climate change, and options for adapting to it, taking into consideration the inter-relationship between vulnerability, adaptation and sustainable development, while the assessed information is considered by sectors (water resources; ecosystems; food & forests; coastal systems; industry; human health) and regions (Africa; Asia; Australia & New Zealand; Europe; Latin America; North America; Polar Regions; Small Islands) [7].

The IPCC Working Group III (WG III) assesses options for mitigating climate change through limiting or preventing greenhouse gas emissions and enhancing activities that remove them from the atmosphere. The main economic sectors taken into account, both in a near-term and in a long-term perspective include energy, transport, buildings, industry, agriculture, forestry, waste management. The WG analyses the costs and benefits of the different approaches to mitigation, considering also the available instruments and policy measures and the approach is more and more solution-oriented [7].

Within the above three working groups, many hundreds of international experts assess climate change in the published Assessment Reports. A landmark year for this kind of reports was the year 2007 when the Fourth Assessment Report was published. This report including the work of the three Working Groups, showed that the need for drastic action to tackle climate change seems to be getting more and more urgent [8, 9, 10].

According to contemporary atmospheric sciences and the 2007 IPCC reports nearly half of the solar radiation is absorbed by the surface of the earth which gets warm. A certain amount of the infrared radiation emitted by the earth's surface is also absorbed and re-emitted in all directions by the greenhouse gas molecules and clouds in the atmosphere. This results in warming the surface of the earth and the lower atmosphere [11]. The mechanism of the Greenhouse Effect is depicted in Figure 1.

Despite the wealth of evidence pointing to the existence of a Global Warming Effect and of human caused Climate Change, there are still 'climate sceptic' scientists and politicians who have certain objections. A fact that is beyond any doubt is the experimental evidence on the time profile of the concentration of greenhouse gases in the atmosphere for the last two millennia. The experimental data on the concentrations of greenhouse gases (Carbon Dioxide: CO_2, Methane: CH_4 and Nitrous Oxide: N_2O) in the atmosphere for the last 2.000 years, show that there is an evident and profound increase in their concentrations after the beginning of the industrial era in the years after 1750 [12]. This is clearly illustrated in Figure 2.

It is estimated that since the year 1750 and onwards anthropogenic or human-caused CO_2 emissions are mainly due to (nearly 2/3rds) fossil fuel combustion for energy and transport (plus a smaller contribution from cement manufacture) and the rest (1/3rd) is due to land use change (primarily deforestation) [13].

In the Summary for Policymakers in the most recent report of the Working Group I of IPCC, the observed changes in the climate system were summarized as follows [14]:

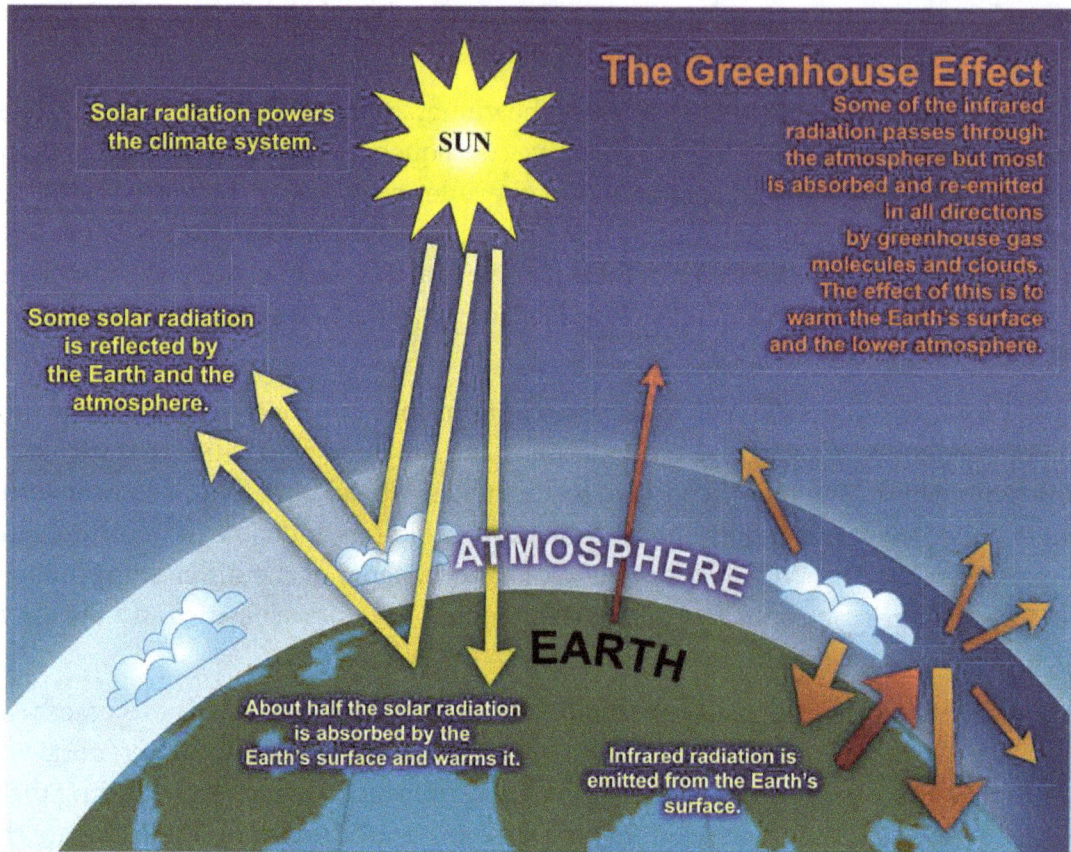

Figure 1. An idealised model of the natural greenhouse effect. From reference 11, p.98. Reproduced by permission from IPCC.

"Warming of the climate system is unequivocal, and since the 1950s, many of the observed changes are unprecedented over decades to millennia. The atmosphere and ocean have warmed, the amounts of snow and ice have diminished, sea level has risen, and the concentrations of greenhouse gases have increased".

In the same document, some rather alarming details of climate change are reported with particular reference to the individual components of the climate system in our planet [14]:

"**Atmosphere**: Each of the last three decades has been successively warmer at the Earth's surface than any preceding decade since 1850. In the Northern Hemisphere, 1983–2012 was likely the warmest 30-year period of the last 1400 years (statement of medium confidence)."

"**Ocean**: Ocean warming dominates the increase in energy stored in the climate system, accounting for more than 90% of the energy accumulated between 1971 and 2010 (statement of high confidence). It is virtually certain that the upper ocean (0–700 m) warmed from 1971 to 2010 (see Figure SPM.3), and it likely warmed between the 1870s and 1971."

"**Cryosphere**: Over the last two decades, the Greenland and Antarctic ice sheets have been losing mass, glaciers have continued to shrink almost worldwide, and Arctic sea ice and Northern Hemisphere spring snow cover have continued to decrease in extent (statement of high confidence)."

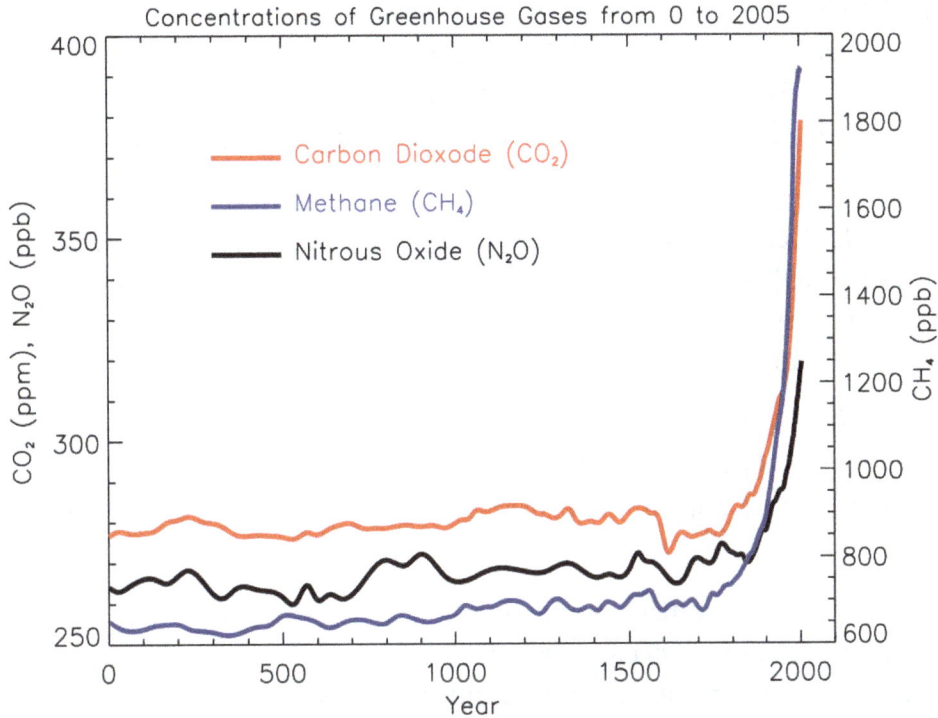

Figure 2. Atmospheric concentrations of important long-lived greenhouse gases over the last 2,000 years. Increases since about 1750 are attributed to human activities in the industrial era. Concentration units are parts per million (ppm) or parts per billion (ppb), indicating the number of molecules of the greenhouse gas per million or billion air molecules, respectively, in an atmospheric sample. From reference 12, p.100. Reproduced by permission from IPCC.

"**Sea Level**: The rate of sea level rise since the mid-19th century has been larger than the mean rate during the previous two millennia (statement of high confidence). Over the period 1901 to 2010, global mean sea level rose by 0.19 [0.17 to 0.21] m."

"**Carbon and Other Biogeochemical Cycles**: The atmospheric concentrations of carbon dioxide, methane, and nitrous oxide have increased to levels unprecedented in at least the last 800,000 years. Carbon dioxide concentrations have increased by 40% since pre-industrial times, primarily from fossil fuel emissions and secondarily from net land use change emissions. The ocean has absorbed about 30% of the emitted anthropogenic carbon dioxide, causing ocean acidification."

The following more detailed observations were reported for the carbon and other biogeo-chemical cycles [14]:

"The atmospheric concentrations of the greenhouse gases carbon dioxide (CO_2), methane (CH_4), and nitrous oxide (N_2O) which have all increased since 1750 due to human activity, in 2011 were 391 ppm, 1803 ppb, and 324 ppb, exceeding their pre-industrial levels by about 40%, 150%, and 20%, respectively. The concentrations of CO_2, CH_4, and N_2O now substantially exceed the highest concentrations recorded in ice cores during the past 800,000 years. The mean rates of increase in atmospheric concentrations over the past century are, with very high confidence, unprecedented in the last 22,000 years."

"The annual CO_2 emissions from fossil fuel combustion and cement production were 8.3 [7.6 to 9.0] GtC·yr^{-1} averaged over 2002–2011 (statement of high confidence) and were 9.5 [8.7 to 10.3] GtC·yr^{-1} in 2011, 54% above the 1990 level. Annual net CO_2 emissions from anthropogenic land use change were 0.9 [0.1 to 1.7] GtC·yr^{-1} on average during 2002 to 2011 (statement of medium confidence)."

"From 1750 to 2011, CO_2 emissions from fossil fuel combustion and cement production have released 375 [345 to 405] GtC to the atmosphere, while deforestation and other land use change are estimated to have released 180 [100 to 260] GtC. This results in cumulative anthropogenic emissions of 555 [470 to 640] GtC."

"Of these cumulative anthropogenic CO_2 emissions, 240 [230 to 250] GtC have accumulated in the atmosphere, 155 [125 to 185] GtC have been taken up by the ocean and 160 [70 to 250] GtC have accumulated in natural terrestrial ecosystems (i.e., the cumulative residual land sink)."

"Ocean acidification is quantified by decreases in pH (pH=$-\log_{10}$ [H$^+$], where [H$^+$] is the concentration of H$^+$). The pH of ocean surface water has decreased by 0.1 since the beginning of the industrial era (statement of high confidence), corresponding to a 26% increase in hydrogen ion concentration"

The general trend resulting in the above increase of the acidity of the oceans is clearly depicted in Figure 3.b.

The curve in Fig. 3.a. is the classic "Keeling Curve". The Keeling Curve is a graph which plots the ongoing change in concentration of carbon dioxide in the atmosphere of the Earth since 1958 [15]. It is based on continuous measurements taken at the Mauna Loa Observatory in Hawaii (and in the South Pole) that began by Charles David Keeling. These measurements were the first significant evidence of rapidly increasing carbon dioxide levels in the atmosphere. The Keeling Curve is considered by many scientists as the graph which brought worldwide attention to the current increase of carbon dioxide in the atmosphere [16].

The Kyoto Protocol was drawn up on 11 December 1997 as an implementary measure to the United Nations Framework Convention on Climate Change (UNFCCC) signed on 9 May 1992 in Rio that set up binding obligations on a number of countries to reduce, below their baselines, emissions of carbon dioxide which are generated mostly by electricity, coal, and steel plants [17]. It is evident from Figure 3.a. that the carbon dioxide emissions have in fact kept increasing with a higher rate after the agreement entailed in the Kyoto Protocol.

The American climatology scientist James Edward Hansen, a pundit of global warming and climate change and a pioneer activist calling for action to mitigate the effects of climate change [18], addressed the reluctance of world leading countries to take drastic measures [19]:

"I have been told by a high government official that I should not talk about "dangerous anthropogenic interference" with climate, because we do not know how much humans are changing the Earth's climate or how much change is "dangerous". Actually, we know quite a lot. Natural regional climate fluctuations remain larger today than human-made effects such

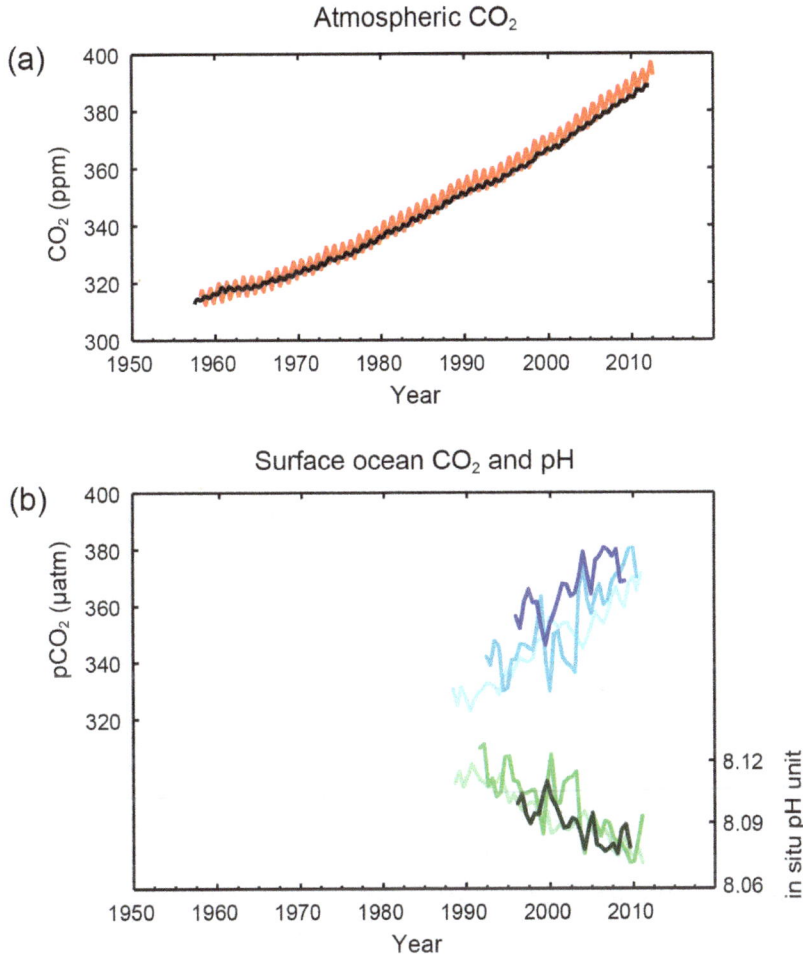

Figure 3. Multiple observed indicators of a changing global carbon cycle: (a) atmospheric concentrations of carbon dioxide (CO_2) from Mauna Loa (19°32′N, 155°34′W – red) and South Pole (89°59′S, 24°48′W – black) since 1958; (b) partial pressure of dissolved CO_2 at the ocean surface (blue curves) and in situ pH (green curves), a measure of the acidity of ocean water. Measurements are from three stations from the Atlantic (29°10′N, 15°30′W – dark blue/dark green; 31°40′N, 64°10′W – blue/green) and the Pacific Oceans (22°45′N, 158°00′W – light blue/light green). Full details of the datasets shown here are provided in the underlying report and the Technical Summary Supplementary Material. {Figures 2.1 and 3.18; Figure TS.5}. From reference 14, p.12. Reproduced by permission from IPCC.

as global warming. But data show that we are at a point where human effects are competing with nature and the balance is shifting."

The argument of James Edward Hansen that we know quite a lot about the "dangerous anthropogenic interference" with climate, is evidenced by his reported work [20, 21, 22]. In their assessment of "dangerous climate change" James Edward Hansen and his colleagues suggested the required reduction of carbon emissions to protect young people, future generations and nature. According to them the carbon emissions should be such that they fulfill a cumulative industrial-era limit of ~500 GtC fossil fuel emissions and 100 GtC storage in the biosphere and soil which would keep climate close to the Holocene range to which humanity and other species are adapted [22]. The urgency of the situation has also been addressed in the most recent report of IPCC [23].

2. Cement chemistry and industry and its impact on global warming

Concrete is a composite material consisting of an aggregate and a binder phase (hardened cement paste) that is formed by the reaction (hydration) of Portland cement clinker with water [24, 25]. This initially rather rapid reaction can continue for many years resulting in an intimate, micro-porous mixture of several crystalline and poorly crystalline phases. The chemical and physical aspects of the structure, as well as certain properties in concrete materials were outlined [26].

Cement represents a commodity which surpasses any other that our species produces and uses as a construction material. It is also true that cement plants are amongst the biggest human made industrial constructions, simply to mention cement rotary kilns with diameter up to 3.7m (12 feet) and length up to 100m (longer in many instances than the height of a 40-story building) [27]. A state of the art diagram showing in the simplest possible way the very complex succession of reactions taking place during the production of the 'grey' cement powder, the so-called cement clinker was reported in Figure 2.6 in page 79 of reference 25.

The estimated annual production of Ordinary Portland Cement (or simply cement) in 2013 was 4 billion metric tonnes with China producing more than half of it (2.3 billion tonnes) followed by India (0.28 billion tonnes) and by USA (0.0778 billion tonnes) [28]. Most of this portland cement is used in making concrete or mortars. Concrete is the most used construction material in the world, with an annual production of around 11 billion metric tonnes [29], despite the fact that it competes in the construction sector with concrete substitutes, such as aluminum, asphalt, clay brick, rammed earth, fiberglass, glass, steel, stone, and wood [28]. The manufacture of concrete is an industry with a worldwide value of 1 trillion US$ which employs nearly 30 million workers [29]. The magnitude of the global cement industry and the use of concrete as a building material are such that on our planet we produce annually nearly 0.55 metric tonnes (550 Kg) of OPC or 1.50 metric tonnes (1,500 Kg) of concrete for each one of the 7.239 billion people living on planet Earth [30].

A general overview on hydraulic (mainly portland) cement and, to some degree, concrete, as well as a description of the monthly and annual U.S. Geological Survey (USGS) cement industry canvasses in general terms of their coverage and some of the issues regarding the collection and interpretation of the data therein was reported [31].

The Ordinary Portland Cement (OPC) clinker production is a complex Solid/Liquid State reaction process. Grinded limestone ($CaCO_3$), Clays (Aluminium Silicates, Al-Si-O), Quartz (SiO_2) and Fe_2O_3 in certain weight ratios are homogeneously mixed in the rotary kiln and heated up to ~1450°C. The quenched product is mainly a mixture of 3 compounds (2 calcium silicates and 1 calcium aluminate) and one solid solution (calcium aluminium ferrate) [24, 32]:

1.   Tricalcium Silicate, **$3CaO \cdot SiO_2$** (Alite), **C_3S**: 45-70 wt.%.

2.   β-Dicalcium Silicate, **$β-2CaO \cdot SiO_2$** (Belite), **$β-C_2S$**: 25-30 wt.%.

3.   Tricalcium Aluminate, **$3CaO \cdot Al_2O_3$** (Celite), **C_3A**: 5-12 wt.%.

4. Tetracalcium Alumino Ferrite (solid solution), **$4CaO \cdot Al_2O_3 \cdot Fe_2O_3$ (Feritte), C_4AF: 5-12 wt. %.**

Using a rather simplified picture of the chemistry involved, the following reactions take place during the production of Ordinary Portland Cement (OPC) clinker:

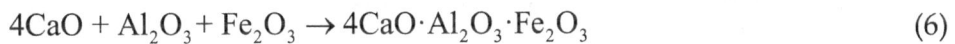

$$CaCO_3 \rightarrow CaO + CO_2 \uparrow \tag{1}$$

$$3CaO + SiO_2 \rightarrow 3CaO \cdot SiO_2 \tag{2}$$

$$2CaO + SiO_2 \rightarrow 2CaO \cdot SiO_2 \tag{3}$$

$$12CaO + 7Al_2O_3 \rightarrow 12CaO \cdot 7Al_2O_3 \tag{4}$$

$$3CaO + Al_2O_3 \rightarrow 3CaO \cdot Al_2O_3 \tag{5}$$

$$4CaO + Al_2O_3 + Fe_2O_3 \rightarrow 4CaO \cdot Al_2O_3 \cdot Fe_2O_3 \tag{6}$$

In the industrial process, Portland cement is manufactured by crushing, milling and proportioning the following materials [33]:

1. Lime or calcium oxide, CaO: from limestone, chalk, shells, shale or calcareous rock.

2. Silica, SiO_2: from sand, old bottles, clay or argillaceous rock.

3. Alumina, Al_2O_3: from bauxite, recycled aluminum, clay.

4. Iron, Fe_2O_3: from clay, iron ore, scrap iron and fly ash.

5. Gypsum, $CaSO_4 \cdot 2H_2O$: found together with limestone.

The materials 1-4 are proportioned to produce a mixture with the desired chemical composition and then ground and blended mainly by a dry process. The materials are then fed through a kiln at 2,600ºF (1427°C) to produce grayish-black pellets known as clinker. The alumina and iron act as fluxing agents which lower the melting point of the mixture from 3,000ºF (1649°C) to 2600ºF (1427°C). After this stage, the clinker is cooled, pulverized and gypsum added to regulate setting time. It is then ground extremely fine to produce cement [33].

The CO_2 footprint of cement production causes more than 5% of the total anthropogenic carbon dioxide (CO_2) emissions with nearly 800 kg CO_2 emissions per 1000 kg of cement produced. These emissions are the process emissions due to the chemical reaction of limestone ($CaCO_3$) decomposition to lime (CaO) (~60%) and the energy-related emissions (~40%) due to the combustion of fossil fuels (to reach the 1450°C reaction temperature) and the power generation needed for grinding the raw materials and the product [34, 35].

$$CaCO_3 \rightarrow CaO + CO_2 \uparrow$$
$$1\,kg \qquad 0.56\,kg + 0.44\,kg$$

(7)

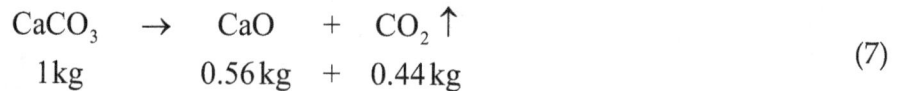

The total CO_2 emissions in the production of cement, including process and energy-related emissions were reviewed [34]. The average intensity of carbon dioxide emissions from total global cement production is 222 Kg of C/1.000 Kg of cement or 814 Kg CO_2/1.000 Kg of cement. Emission mitigation options include energy efficiency improvement, new processes, a shift to low carbon fuels, application of waste fuels, increased use of additives in cement making, and, eventually, alternative cements and CO_2 removal from flue gases in clinker kilns [34 and references there in].

According to the America's Cement Manufacturers Portland Cement Association (PCA) Voluntary Code of Conduct, within the Cement Manufacturing Sustainability (CMS) Program, the Environmental Performance Measures translate the Cement Manufacturing Sustainability (CMS) Program Principles into action. Currently there are four goals approved by the PCA members [36]:

1. **Carbon Dioxide (CO_2)** – The U.S. cement industry has adopted a year 2020 voluntary target of reducing CO_2 emissions by 10 percent (from a 1990 baseline) per ton of cementitious product produced or sold.

2. **Cement Kiln Dust (CKD)** – The U.S. cement industry has adopted a year 2020 voluntary target of a 60 percent reduction (from a 1990 baseline) in the amount of CKD landfilled per ton of clinker produced.

3. **Environmental Management Systems (EMS)** – The U.S. cement industry has adopted a year 2006 voluntary target of at least 40 percent implementation of U.S. cement plants using an auditable and verifiable EMS with 75 percent of the U.S. plants implementing an EMS by the end of 2010, and with 90 percent by the end of 2020.

4. **Energy Efficiency** – The U.S. cement industry has adopted a year 2020 voluntary target of 20 percent improvement (from 1990 baseline) in energy efficiency – as measured by total Btu-equivalent per unit of cementitious product.

It is apparent that the target of 10% reduction by year 2020 in Carbon Dioxide (CO_2) emissions from the USA cement industry represents a rather marginal evolutionary approach in tackling the impact of cement industry in global warming. It also stresses the need for development of alternative new technologies and processes for the cement manufacture.

3. Thermodynamic aspects of the synthesis of Ordinary Portland Cement (OPC) clinker with emphasis on calcium silicates and calcium aluminates

The formation of Ordinary Portland Cement (OPC) clinker is a very complex process with reactions involving at least the four oxides CaO, SiO_2, Al_2O_3 and Fe_2O_3. It is generally accepted that the knowledge of the following phase diagrams is very important for the understanding of the cement production process, especially those involving CaO, SiO_2 and Al_2O_3:

i.   Binary phase diagrams of $CaO-SiO_2$ and $CaO-Al_2O_3$.

ii.   Ternary phase diagram of $CaO-SiO_2-Al_2O_3$.

iii.   Quaternary phase diagram of $CaO-SiO_2-Al_2O_3-Fe_2O_3$.

All the above phase diagrams were thoroughly reported and assessed in the Slag Atlas [37 and references cited there in].

The existing phase diagrams of the $CaO-SiO_2$ and of the $CaO-Al_2O_3$ binary systems [37 and references cited there in] show that the very high melting oxide CaO (m.pt. 2613°C for lime) reacts with SiO_2 (m.pt. 1713°C for vitreous and 1722°C for the cristobalite phase) and with Al_2O_3 (2054°C for corundum phase) [melting points are from reference 38] yielding the following major cement compounds:

1. $3CaO \cdot SiO_2$ / Ca_3SiO_5 / C_3S (Mineral Name: Hatrurite) [39] (Cement Chemistry Name: Alite), which melts incongruently at 2150°C and decomposes to CaO and C_2S below 1250°C [37] if not quenched fast, as is the standard practice in the cement industry.

2. $\beta-2CaO \cdot SiO_2$ / $\beta-Ca_2SiO_4$ / $\beta-C_2S$ (Mineral Name: Larnite) [40] (Cement Chemistry Name: Belite), which melts at 2130°C [37] and is transformed to $\gamma-C_2S$ below 500°C if not quenched fast, as is the standard practice in the cement industry.

3. $3CaO \cdot Al_2O_3$ / $Ca_3Al_2O_6$ / C_3A, which melts incongruently at 1539°C and is transformed to another polymorph below 1400°C [37].

4. $12CaO \cdot 7Al_2O_3$ / $Ca_{12}Al_{14}O_{33}$ / $C_{12}A_7$ which is an intermediate compound (Mineral Name: Mayenite) [41] during the synthesis of C_3A, the $C_{12}A_7$ melts at 1413°C and is transformed to another polymorph below1400°C [37].

In the current industrial process of production of Ordinary Portland Cement (OPC) clinker an important step is the fast cooling of the clinker product from temperatures around 1500°C to temperatures well below 1000°C. This quenching of the very high temperature clinker product is necessary in order to prevent the decomposition of alite $3CaO \cdot SiO_2$ (C_3S) to $2CaO \cdot SiO_2$ (C_2S) and CaO (the decomposition of C_3S is taking place below 1250°C) and to avoid the transformation of belite $\beta-2CaO \cdot SiO_2$ ($\beta-C_2S$) (the desired cement phase of C_2S is the belite $\beta-C_2S$) to $\gamma-2CaO \cdot SiO_2$ ($\gamma-C_2S$), since below 500°C and especially at room temperature the stable phase of C_2S is not $\beta-C_2S$ but $\gamma-C_2S$.

Again it is important to emphasize that the raw meal used in the industrial production of Ordinary Portland Cement (OPC) clinker is composed not only by the 'ideal' $CaCO_3$, SiO_2, Al_2O_3 and Fe_2O_3 reactants but also there are a lot of other cations and anions present. All these ions and the precise temperature profile of the reactions lead to the formation of more than five polymorphs for the quenched $3CaO \cdot SiO_2$ (C_3S) and the presence of some $\gamma-C_2S$ along with the desired $\beta-C_2S$ for the $2CaO \cdot SiO_2$ (C_2S).

The high temperature reactions taking place in the rotary kiln during cement production and the complexity of the processes involved require a detailed knowledge of these reactions in order to achieve technological advances. An introduction to the topic of cement, including an overview of cement production, selected cement properties, and clinker phase relations was reported [42]. The authors provided an extended summary of laboratory-scale investigations on clinkerization reactions and on the most important reactions in cement production. The formation of clinker by solid state reactions, solid–liquid and liquid–liquid reactions, as well as the influences of particles sizes on clinker phase formation was discussed. In addition, a mechanism for clinker phase formation in an industrial rotary kiln reactor was outlined [42].

In the quest to understand the fundamentals of the cooling process of cement clinker in order to improve the ongoing cooling process, the decomposition of alite (C_3S) in Portland cement clinker was investigated by isothermal annealing [43]. The clinker phases were analyzed by quantitative X-ray diffraction and scanning electron microscopy in order to investigate the microstructure. It was found that the fastest decomposition rate of Ca_3SiO_5 (C_3S) to Ca_2SiO_4 (C_2S) and CaO appeared at 1125–1150°C using a temperature–time–transformation diagram. The combined XRD and SEM study showed that the decomposition of alite (Ca_3SiO_5) primarily occurred at the cracks, edges, and defects of the clinker, while the resulting free CaO segregated and subsequently controlled the decomposition rate of alite. The decomposition kinetics of alite was described by a three-dimensional Jander diffusion model with a non-Arrhenius behavior for the activation energy which exhibited two distinct linear functions with temperature, one above (higher activation energies) and the other below (lower activation energies) the temperature of 1380.65 K (1107.49°C). It was reported that interstitial phases recrystallized during the annealing process were accompanied by an increase of the $3CaO \cdot Al_2O_3$ (C_3A) and $4CaO \cdot Al_2O_3 \cdot Fe_2O_3$ (C_4AF) contents and that the recrystallization of C_3A was temperature-dependent, especially above 1000°C [43].

4. Synthesis of Ordinary Portland Cement (OPC) clinker compounds by various methods with emphasis on calcium silicates

A number of methods used in the synthesis of calcium silicates and calcium aluminates, include solid-state reactions, growth from the melt, combustion and sol gel synthesis as well as other more special synthetic methods. The energy characteristics of these synthetic techniques can be compared with the molten salt synthesis used in previous studies and in the present work. A straightforward means of comparison of the molten salt synthesis used in this work with other methods of synthesis involves mainly the reaction temperature, the specific kind of chemical compounds reacting, as well as the kind of reactors used in each method.

4.1. Solid state synthesis of calcium silicates

In a typical solid-state synthesis the two reactants are thoroughly ground in order to increase their reactivity and subsequently are homogeneously mixed in order to increase their contact. The restrictions imposed by the limited mobility of the reacting species as compared to reactions in solution, dictate the use of rather very high temperatures in order to drive the reaction to the desired product. The nucleation and growth of the product as well as the diffusion of one or both reactants are particularly important factors that affect the outcome of the solid-state reactions. The solid-solid reaction rate and the role of the contact points between particles on the reactivity of solids was studied [44]. Taking into account the exact number of contact points between the two components in a system of solids, their reactivity was expressed in terms of the molar ratio, the particle size ratio and the nature of the reacting system. The theory with a particular emphasis on the Jander model was confirmed by existing experimental data, revealing that the contact points play an important role for wide ranges of reactivity of solids [44]. The basics and mathematical fundamentals of solid-state kinetic models were reviewed [45]. Models used in solid-state kinetic studies were classified according to their mechanistic basis as nucleation, geometrical contraction, diffusion, and reaction order. The authors summarized commonly employed models and presented their mathematical development [45]. There is a wealth of information on the synthesis of calcium silicates by reaction in the solid state done in Europe mainly in Germany more than five decades ago [46, 47]. The formation of the orthosilicate β-$2CaO \cdot SiO_2$, Wollastonite $CaO \cdot SiO_2$ and $3CaO \cdot 2SiO_2$ by solid-state reaction of CaO with H_4SiO_4 (silicic acid) in temperatures of 1000°C and for up to 400 hours was examined by microscopy and XRD [46]. A detailed XRD and DTA study of the solid-state synthesis of calcium silicates (CS, C_2S, C_3S) by the reaction of $CaCO_3$ or $Ca(OH)_2$ with amorphous SiO_2 at various temperatures up to 1500°C for 3 hours was reported and a scheme of reaction of CaO with SiO_2 was proposed [47]. The kinetics of the thermal synthesis of calcium silicates by the dynamic reaction of CaO with SiO_2 (2:1 mole ratio) was studied at temperatures up to 1500°C (especially above 1400°C) by simultaneous TG, DTA, high-temperature X-ray diffraction (horizontal sample) and microscopy as well as by dilatometry. The reaction mechanism in the dynamic thermal synthesis of Ca_2SiO_4 was elucidated and confirmed by high-temperature X-ray diffraction [48]. A quantitative kinetics study showed that the reaction between CaO and a SiO_2-rich liquid phase is the rate-determining step. The calculated from the kinetic data activation parameters agree well with the proposed reaction mechanism [49]. The preparation method of β-C_2S powders without stabilizer and their hydration characteristics were studied [50]. The β-C_2S was formed when γ-C_2S was heated at about 1000°C or at even higher temperatures of about 1500°C. The hydration kinetics of β-C_2S produced from α'-C_2S was found to be markedly different from that produced from α-C_2S, with very small amount of $Ca(OH)_2$ produced in the hydration of β-C_2S without stabilizer and the formed C-S-H had a composition of C/S~2 [50]. The composition of phases in the reaction zones of the CaO-SiO_2 binary system was determined by electron probe microanalysis in the temperature range between 1000 – 1450 °C and the dependence on the time of annealing was investigated [51]. Utilising the diffusion couple technique in combination with a new designed screw loaded sample holder it was suggested that the growth of all product layers obeys the parabolic rate law for diffusion-controlled solid state reactions. The respective reaction rate constants and activation energies were presented and discussed [51]. The solid state synthesis of pure Portland cement phases was reported [52]. The pure clinker compounds are often used to test various aspects of cement chemistry in particular hydration behaviour, mainly due to the fact that they are nearly isostructural to the Ordinary Portland Cement (OPC) clinker phases. The apparent need of efficient methods to produce big quantities of the pure phases with an affordable cost is well documented by the high prices of these pure phases available from suppliers of chemicals. The authors reported the synthesis of the pure phases as well as a description of phase relations and possible polymorphs of the four main phases in Portland cement, i.e. tricalcium silicate, dicalcium silicate, tricalcium aluminate and tetracalcium alumino ferrite. In addition, details of the process of solid state synthesis were described including state of the art practical advice on equipment and techniques [52]. It is obvious from all the above studies that the solid state synthesis of calcium silicates involves reactions in temperatures well above 1000°C and often up to 1500°C.

4.2. Growth from the melt synthesis of calcium silicates

The same range of temperatures (1000°C up to 1500°C) seem to be used in the growth from the melt synthesis of calcium silicates. A sample holding technique consisting of a thin platinum wire loop was used for the study of crystal growth in silicate melts (SiO_2 ~49-51 wt. %, Al_2O_3 ~10-15 wt.%, FeO ~9-19 wt.%, MgO ~9 wt.%, CaO ~10-11 wt.%, Na_2O ~0.05-3 wt.%,) in gas mixing furnace. The experiments reported were run from 1 hour up to 24 hours at 1250°C [53]. A novel technique of dynamic crystallization used for the directional devitrification from a molten zone have been described by Maries and Rogers [54]. Using this technique it was possible to prepare aligned fibrous crystals by drawing filaments of glass through a supported molten zone. The researchers subsequently modified the method in order the filament to be drawn from a melt through an orifice in the base of a conical platinum crucible [55]. Using this process they crystallized β-$CaSiO_3$ filaments by continuous drawing from a melt contained in a resistance-heated platinum crucible, whereby the temperature of crystallization was determined primarily by the speed of drawing. The temperatures used were apparently above the 1544°C melting point of $CaO \cdot SiO_2$. The authors have also used this technique for the preparation of filaments consisting of aligned fibres of fluoramphiboles, which are synthetic analogues of natural amphiboles by isomorphous substitution of hydroxyl by fluorine anions. The method led to the synthesis of aligned fibrous crystals which bear resemblance with natural amphibole asbestos [56].

The morphology of calcium metasilicate $CaSiO_3$ produced during the crystallization of glasses and melts of approximately metasilicate composition has been investigated employing isothermal heat treatments and a dynamic crystal-pulling technique [57]. It was observed that the crystallization took place by a dendritic or spherulitic mechanism, depending on which of the crystal polymorphs is stable under the prevailing conditions. It was found that the morphology of the crystals is controlled by the ease with which the anionic groups present in the amorphous phase can be incorporated into the growing crystals. Time-temperature-transformation diagrams have been constructed from the experimental data. It should be noted that the glass compositions used to investigate the crystallization characteristics of calcium metasilicate contained not only CaO (32.5-37.4 mole %) and SiO_2 (51.2-55.4 mole %) but also other oxides such as Al_2O_3 (2.6-7.4 mole %), ZnO (5.7-15.4 mole %) and Na_2O (4.2-4.8 mole %) and in one case a fluoride CaF_2 (5.3 mole %). It should be emphasized that due to the very high melting points of certain compositions, it was necessary to use very high temperatures, well above 1000°C, i.e. time-temperature-transformation diagram for a certain composition rods nucleated on a β-CaSiO₃ block and on a Pt/Pt 13% Rh thermocouple were reported from T_F=1340°C to 20°C [57].

The rather very high temperatures used in both the solid state synthesis reactions and in the growth from the melt synthesis show the importance of molten salt synthesis for the preparation of calcium silicates in temperatures quite lower than the 1500°C.

4.3. Combustion synthesis of calcium silicates

The low-temperature combustion synthesis of nanocrystalline β-dicalcium silicate (β-Ca_2SiO_4) with high specific surface area was reported [58]. The synthesis of β-Ca_2SiO_4 was achieved for the first time by a simple solution combustion method using citric acid, $C_6H_8O_7$ (semi-structural formula: $HOOCCH_2C(OH)(COOH)CH_2COOH$) as fuel. The effect of calcination temperature on the average crystallite size, specific surface area and morphology of the powders were studied by X-ray diffraction (XRD), scanning electron microscopy (SEM) and N_2 adsorption measurements (BET). It was found that when increasing the calcination temperature of β-Ca_2SiO_4 from 650°C to 1100°C this results in the production of crystallites of larger size and lower specific surface area of β-Ca_2SiO_4. It was shown that the highest specific surface area measured was up to 26.7 m² g⁻¹ when the powders were calcined at 650°C [58]. The use of the combustion method for the synthesis of calcium silicates looks very interesting for specific applications of the product which can make affordable the cost of the method.

4.4. Sol gel synthesis of calcium silicates

The synthesis of calcium silicates by a chelate gel route using aqueous solution of citric acid was reported [59]. A number of gel techniques such as the metal-chelate gel method, in situ polymerized complex method and polymer precursor method used in the preparation of ceramics have the potential to yield compositionally homogeneous and fine powders with a narrow particle size distribution [59]. The presence of the various polymorphs and the particle size distribution are particularly important factors since they affect the hydration activity of calcium silicates. A metal-chelate gel route based on gelation of the aqueous solution of citric acid has been successfully applied to the synthesis of calcium silicates (Ca_2SiO_4 and Ca_3SiO_5) for the first time and their phase transformations and particle size were discussed in comparison to the conventional solid-state reaction route [59]. It was found that the novel citrate gel route produces β-Ca_2SiO_4 (the high-temperature phase and favourable cement compound) while the conventional solid-state reaction route produces γ-Ca_2SiO_4 (the low-temperature phase) This result was interpreted in terms of the particle size effect and the energy barrier. It was suggested that the nucleation and propagation of microcracks result in overcoming a comparatively high-energy barrier, $\Delta G^*(\beta \rightarrow \gamma)$ and that the particle size effect governs both the statistic of martensitic nucleation and the propagation of the $\beta \rightarrow \gamma$ transformation. In the case of tricalcium silicate, the triclinic Ca_3SiO_5 (the low-temperature phase) was produced by both the citrate gel route and the conventional solid-state reaction route. It was suggested that the nucleation and propagation do not result in the M-T transformation, pointing that the energy barrier of the monoclinic (M) to the triclinic (T) transformation, $\Delta G^*(M\text{-}T)$ is rather small [59].

The sol–gel synthesis and the textural characterisation of mesoporous calcium silicate glasses having compositions within the liquid–liquid immiscibility dome of the CaO–SiO_2 system was reported [60, 61].

A number of crack-free silica–calcia xerogel monoliths of various shapes and sizes and having compositions of $xCaO \cdot (1-x)SiO_2$ ($0 \leq x \leq 0.5$, x=mole fraction), were prepared via the sol–gel processing technique, using tetraethyl orthosilicate, $Si(OC_2H_5)_4$ (TEOS) and calcium nitrate, $Ca(NO_3)_2$ reactants [60]. The homogeneous throughout the monolith gel-glasses were characterised by X-ray diffraction, infrared (FTIR) spectroscopy, energy dispersive scanning electron microscopy (SEM-EDS) and by differential thermal analysis (DTA). It was found that they are amorphous even after stabilisation at 600°C, they have crystallisation temperatures above 850°C and they formed crystalline phases present in the CaO–SiO_2 phase diagram only when sintered at 1000°C [60].

The textural characteristics including the surface area and porosity (pore structure and pore volume), of the above monolith gel-glasses were studied by nitrogen adsorption, mercury porosimetry and helium pycnometry [61]. The nitrogen adsorption and the mercury porosimetry methods which study the pore morphology showed that the pore system in these gel-glasses consists of a three-dimensional network of cavities (pores) interconnected by constrictions (throats) in the mesopore range (75–314 Å in diameter) with the pore sizes dependent on the composition of the gel-glasses. It was found that when the CaO content was decreased, then the surface area increased, the pore size decreased, while the skeletal and bulk densities both increased and the gel-glasses were of approximately 30% porosity. [61].

The synthesis of pure cementitious phases by sol-gel process as precursor was reported [62]. The pure phases of calcium silicates and aluminates which are the main constituents of ordinary Portland cement (OPC) and calcium aluminate cements (CAC) are of great importance for the cement research. Due to the demand for big amounts of these pure phases and the fact that their synthesis by solid-state reactions is difficult, there is an obvious need for more efficient synthetic methods. The authors suggested that an attractive alternative to the conventional synthetic route is the sol-gel process. Experimental results on the improved synthesis of calcium silicates and aluminates via sol-gel processes were reported, along with the characterization of the pure clinker phases and studies of their hydration behaviour [62].

The sol–gel synthesis of belite (β-Ca_2SiO_4), one of the major compounds in Portland cement clinker was reported [63]. The authors emphasized the fact that in the conventional preparation with solid state reaction, belite is produced by long lasting sintering of limestone and quartz at temperatures exceeding 1400°C. In the sol–gel synthesis of belite reported, both aqueous and non-aqueous sol–gel routes were applied and the preparation of the precursor mixture and the formation of the ceramic product were monitored using TG/

DTG, XRD, FT-IR and SEM. The combined use of the above techniques led to the recording of all the transformations occurring during the processing of the precursors and the formation of the final products. It was found that both the aqueous and the non-aqueous sol–gel routes can be successful in the preparation of di-calcium silicate and that the final products consisted of very fine spherical crystallites with size in the range 1–3 μm, whose formation required a 3 hours sintering at 1000°C, but in both cases β-Ca$_2$SiO$_4$ was obtained without the use of any chemical stabilizers [63].

The sol-gel synthesis of calcium silicates offer a lot of opportunities by controlling the preparation and giving access to tailor made products with specific particle size, shape and morphology. It can be particularly attractive for the synthesis of high price cements for very special applications.

4.5. Other methods of synthesis of calcium silicates

Other methods of synthesis of calcium silicates include rather special techniques such as hydrothermal synthesis, the Pechini method, Ultrasonical – Sonochemical techniques, as well as the so called Organic Steric Entrapment method.

The hydrothermal methods initially produce in a temperature of a just a few hundreds of °C degrees some kind of calcium silicate hydrates, which have to be thermally processed in much higher temperature in order to produce the anhydrous calcium silicates. An example is the microwave-assisted hydrothermal preparation of Ca$_6$Si$_6$O$_{17}$(OH)$_2$ and of β-CaSiO$_3$ nanobelts [64]. The initial step involved xonotlite (Ca$_6$Si$_6$O$_{17}$(OH)$_2$) nanobelts which were synthesized by a microwave-assisted hydrothermal method at 180°C for 90 minutes using a feeding molar ratio of Ca(NO$_3$)$_2$·4H$_2$O to Na$_2$SiO$_3$·9H$_2$O in the range of 0.8–3.0. The crystalline wollastonite (β-CaSiO$_3$) nanobelts were obtained by microwave thermal transformation at 800°C for 2 h of Ca$_6$Si$_6$O$_{17}$(OH)$_2$ nanobelts which were used as both the precursor and the template for the preparation of β-CaSiO$_3$ nanobelts [64].

In the Pechini process the pyrolysis of the polymer matrix leads to the development of the oxide precursor which has a high degree of homogeneity [65]. A relevant study reported involved the hydration kinetics and the phase stability of reactive dicalcium silicate synthesized by the Pechini process [66]. It was found that when increasing the calcination temperature, the amorphous product first crystallized to α'$_L$-phase and subsequently to the ß-and γ-phases. It was also reported that the specific surface area, ranging from 40 to 1 m^2/g, strongly depended on the calcination temperature of 700°-1200°C for 1 hour [66].

In the ultrasonic or sonochemical methods at some stage of the process the reaction mixture is sonicated in an ultrasonic bath. A relevant work was on the synthesis and characterization of hydration behaviour of manganese oxide-doped dicalcium silicates obtained from rice hull ash [67]. In this work the authors reported that the syntheses were performed using silica obtained from rice hull ash and the solids SiO$_2$, CaO and MnO were weighed in stoichiometric proportions to prepare silicates having a ratio (Ca+Mn)/Si=2, with the amount of manganese varying from 1 to 10 mole %. The ground solid reactants after the addition of water formed aqueous suspensions which were sonicated for 60 minutes in an ultrasonic bath. This was followed by drying, and the resulting solids were ground and burned at 800°C producing calcium silicates containing up to 10% of manganese oxide [67].

The synthesis and hydration study of Portland cement components prepared by the organic steric entrapment method was reported [68]. The major four components of Portland cement; dicalcium silicate (Ca_2SiO_4), tricalcium silicate (Ca_3SiO_5), tricalcium aluminate ($Ca_3Al_2O_6$), and tetracalcium aluminate iron oxide ($Ca_4Al_2Fe_3O_{10}$) were synthesized by the PVA, $[CH_2CH(OH)]_n$ (Polyvinyl Alcohol) complexation process. The authors stated that powders prepared by this new method can make relatively high yields of pure, synthetic, cement components of nano or sub-micron crystallite dimensions, high specific surface area, and extremely high reactivity at relatively low calcination temperatures in comparison with conventional methods. It is obvious that the above advantages can enhance setting speed, increase strength, and lead to other desirable characteristics of Portland cement [68].

All the above special methods of synthesis yield calcium silicates with very specific properties and are indeed offering a great deal of control over the outcome of the reaction. They are the choice to be considered for special applications that can pay off their production cost. In the question if these special methods can be used in the manufacture of cement on an industrial scale, an obvious answer is that this will only be possible if their cost per unit volume of product becomes comparable to the cost of cement produced with the current industrial process.

4.6. Previous studies on the molten salt synthesis of calcium silicates

Reactions in molten salt media [69] have been used in the synthesis of refractory [70], nuclear [71, 72] and advanced engineering materials [73] as well as in the preparation of low-dimensional solid-state compounds [74] and of semiconductors [75]. In recent years nanomaterials are prepared by molten salt reaction routes [76, 77]. Previously reported successful preparations of dicalcium silicate in molten salt media include the following:

1. Growth of Ca_2SiO_4 crystals in extremely fine (0.01 x 0.1 µm; surface area 120 m²/g) acicular form from the reaction of $CaCl_2$ flux with SiO_2 at 1000°C for 2 hours and at 1200°C for 1 hour [78].

2. Growth of prismatic Ca_2SiO_4 crystals with edge lengths up to 20 mm from a $CaCl_2$ flux [79, 80, 81].

3. Formation of needle-like β-Ca_2SiO_4 crystals upon melting a CaO-SiO_2 mixture in an alkali-halides flux [82]. The crystal growth of Monticellite ($CaMgSiO_4$) and Akermanite ($Ca_2MgSi_2O_7$) using alkali chlorides was also reported [83].

4. Growth of Mn-doped mono-crystals of Ca_2SiO_4 in a $CaCl_2$ flux at 1060-1300°C using raw mixes ($2CaCO_3$, SiO_2, $3CaCl_2$) with different contents of $MnCO_3$ and 1% $Ca_3(PO_4)_2$ equivalent to 0.46% P_2O_5 stabilizer [84].

The growth of Ca_2SiO_4 crystals [85] and $Ca_3Si_2O_7$ $1/3CaCl_2$ single crystals [86] from fluxes were reported. The crystal growth of calcium silicates from melts, hydrothermal solutions and from fluxes was reviewed by Hermoneit and Ziemer [87].

5. The use of low carbon low energy molten salt synthesis method to prepare cement compounds

5.1. The `ULECeS': Ultra low energy cement synthesis: A radical process change to achieve green and sustainable technologies EPSRC EP/F014449/1 project in University College London. Main targets and summary of results

The basic concept of `ULECeS' project, that is the idea of cement production using synthesis in molten salt solvents, was conceived by Alan Maries, following the work on 'Making ceramic powders from molten salts' of Douglas Inman and his research team in 1996. The focus of the study of low carbon low energy molten salt synthesis of cements within the `ULECeS' project in UCL was mainly the energy-related emissions of the cement industry. The `ULECeS' project involved research in new processes in molten salt media for the Ultra Low Energy Cement Synthesis. The target was to attempt a revolutionary rather than an evolutionary approach in improving the environmental impact of the cement industry. Experiments were undertaken to produce cement and cement compounds in molten salt media at much lower reaction temperatures than the 1450°C of the currently used worldwide process.

The `ULECeS': Ultra Low Energy Cement Synthesis: A Radical Process Change to Achieve Green and Sustainable Technologies EPSRC EP/F014449/1 project had the following targets:

1. Synthesis of cements/cement compounds (Ca_3SiO_5, β-Ca_2SiO_4, $Ca_{12}Al_7O_{33}$, $Ca_3Al_2O_6$) in molten salt solvents at lower temperatures than the current industrial process.

2. Preparation of cements/cement compounds with desired particle size distribution, which would reduce the energy needed for grinding the product.

3. The current 'fossil fuel' energy powered process involves a lot of waste of energy. If the molten salt synthesis of cements/cement compounds (Ca_3SiO_5, β-Ca_2SiO_4, $Ca_{12}Al_7O_{33}$, $Ca_3Al_2O_6$) is achieved in lower temperatures (e.g. below 1000°C) then this will pave the way to potential 'electric' energy powered cement production processes.

Experiments on the synthesis of the major cement compounds **Tricalcium Silicate** (Ca_3SiO_5, **C₃S**), **β-Dicalcium Silicate** (β-Ca_2SiO_4, **β-C₂S**) and **Tricalcium Aluminate** ($3CaO \cdot Al_2O_3$, **C₃A**) were attempted in molten alkali chloride (ACl, A=Na, K, Na-K binary eutectic) and $CaCl_2$ solvents. To simulate the real life industrial process the reactants used were $CaCO_3$ (calcite), SiO_2 (α-quartz), Al_2O_3 (corundum) and $NaAlO_2$ ($Na_2O \cdot Al_2O_3$). The thermodynamics of the formation of calcium silicates and aluminates, made it necessary to use solvents with melting points higher than 700°C, which at the same time are not reactive towards the desired product(s). The above prerequisites pointed to the use of alkali and alkaline earth chlorides.

In the next section the results on the attempted molten salt synthesis of the calcium silicates, namely **Tricalcium Silicate** (Ca_3SiO_5, **C₃S**) and **β-Dicalcium Silicate** (β-Ca_2SiO_4, **β-C₂S**) in molten NaCl will be reviewed from the published work on the raw reaction products [88]. The published work on the attempted molten salt synthesis of dicalcium silicate and of tricalcium silicate probed mainly by powder XRD and reported in reference 88, was on the raw 'unpro-cessed' reaction products. New results of processing the raw reaction products will be reported here and compared with the results published in reference 88 on the raw reaction mixes. It is worth mentioning that within the ULECeS project attempts were made to prepare calcium aluminates in molten salt media, with emphasis on Tricalcium Aluminate. This represents approximately the 10 wt% of Ordinary Portland Cement clinker and considerably more of calcium aluminate cements. In all attempts to prepare **Tricalcium Aluminate** ($3CaO \cdot Al_2O_3$, C_3A) by the reaction of $CaCO_3$ with Al_2O_3 in molten alkali chloride solvents at temperatures from 900°C up to 1140°C, the main product was **Dodecacalcium Heptaaluminate** ($12CaO \cdot 7Al_2O_3$, $C_{12}A_7$) which is an intermediate compound apparently reacting with CaO well above 1140°C to produce **Tricalcium Aluminate** ($3CaO \cdot Al_2O_3$, C_3A) [89, 90].

5.2. Studies of low carbon low energy molten salt synthesis of calcium silicates cement compounds and the effect of processing on the reaction product

In this section the published results [88] on the raw 'unprocessed' reaction product of the molten salt synthesis of calcium silicates will be reviewed and compared with new results on the 'processed' reaction product. The processing of the samples involved the removal of the NaCl molten salt solvent by two methods. The first method involved the use of water to remove the NaCl molten salt solvent from the raw reaction product using filtration of the aqueous solution. The second method involved the use of a high temperature filtration technique called 'cupellation', suggested by Douglas Inman. The details of the experimental procedures about the chemicals used, the sample preparation method, and the characterization techniques (Raman spectroscopy / microscopy, powder XRD and SEM) can be found in reference 88.

The study of the low carbon low energy synthesis of calcium silicates involved experiments on the attempted molten salt synthesis of Dicalcium Silicate Ca_2SiO_4 (C_2S) and of Tricalcium Silicate Ca_3SiO_5 (C_3S) by the reaction of $CaCO_3$ (calcite) with SiO_2 (α-quartz) (mole ratios 2:1 and 3:1 for the production of C_2S and C_3S respectively) in molten NaCl and at temperatures from 908°C up to 1140°C using different concentrations of the reactants in the molten salt solvent (Table 1) [88]. The following reactions take place during the formastion of C_2S and of C_3S respectively:

$$2CaCO_3 + SiO_2 \rightarrow Ca_2SiO_4 + 2CO_2 \uparrow \qquad (8)$$

$$3CaCO_3 + SiO_2 \rightarrow Ca_3SiO_5 + 3CO_2 \uparrow \qquad (9)$$

The raw reaction products were characterized by Powder X-Ray Diffraction, Raman Spectroscopy and Scanning Electron Microscopy (SEM). It was reported that in all experiments the major product was β-Ca_2SiO_4 (β-C_2S) even in the case when the intended product was Ca_3SiO_5 (C_3S) [88]. In these samples intended to produce Ca_3SiO_5 (C_3S) a certain amount of CaO that has not reacted with SiO_2 was always detected in the reaction mixture. This finding pointed the need of using temperatures definitely higher than the maximum temperature of 1140°C used in this study in order to produce Ca_3SiO_5 (C_3S) [25]. The reaction of β-Ca_2SiO_4 with CaO is a classic Lux-Flood type Acid-Base reaction:

Sample number/target compound	$CaCO_3/SiO_2/NaCl$ mole ratios	Reaction temperature/time	Reaction products detected by XRD
1.A/β-Ca$_2$SiO$_4$	2.0:1.0:19.2	908°C/2h	β-Ca$_2$SiO$_4$, CaO
1.B/ β-Ca$_2$SiO$_4$	2.0:1.0:19.2	908°C/2h	β-Ca$_2$SiO$_4$, CaO, CaCO$_3$*
2.A/Ca$_3$SiO$_5$	3.0:1.0:19.8	908°C/2h	β-Ca$_2$SiO$_4$, CaO
2.B/Ca$_3$SiO$_5$	3.0:1.0:19.8	908°C/2h	CaO, Ca$_5$(SiO$_4$)$_2$CO$_3$*
2.C/Ca$_3$SiO$_5$	3.0:1.0:19.8	1000°C/1h	β-Ca$_2$SiO$_4$, CaO, trace Ca$_3$SiO$_5$
3 /β-Ca$_2$SiO$_4$	2.0:1.0:20.4	1140°C/3h	β-Ca$_2$SiO$_4$
4 /β-Ca$_2$SiO$_4$	2.0:1.0:13.5	1140°C/3h	β-Ca$_2$SiO$_4$
5 /β-Ca$_2$SiO$_4$	2.0:1.0:10.3	1140°C/3h	β-Ca$_2$SiO$_4$
6 /β-Ca$_2$SiO$_4$	2.0:1.0: 8.0	1140°C/3h	β-Ca$_2$SiO$_4$
7 /Ca$_3$SiO$_5$	3.0-1.0:20.0	1140°C/3h	β-Ca$_2$SiO$_4$,CaO, trace Ca$_3$SiO$_5$
8 /Ca$_3$SiO$_5$	3.0:1.0:14.0	1140°C/3h	β-Ca$_2$SiO$_4$, CaO, trace Ca$_3$SiO$_5$
9 /Ca$_3$SiO$_5$	3.0:1.0: 9.9	1140°C/3h	β-Ca$_2$SiO$_4$, CaO, trace Ca$_3$SiO$_5$
10 /Ca$_3$SiO$_5$	3.0:1.0: 8.1	1140°C/3h	β-Ca$_2$SiO$_4$, CaO, trace Ca$_3$SiO$_5$

*Reaction product in the upper part of the sample crucible only.

Table 1. Powder XRD results for light green grain raw product of Molten Salt Synthesis of Calcium Silicates. Republished with permission of Maney Publishing, from Advances in Applied Ceramics, G. Photiadis et al., 110, 3, 2011 [Reference 88].

$$Ca_2SiO_4 \quad + \quad CaO \quad \rightarrow \quad Ca_3SiO_5$$
$$\text{Acid} \qquad\qquad \text{Base} \qquad\qquad\qquad\qquad \text{(10)}$$

The outcome of the reactions was fully confirmed by Raman spectroscopy which showed the absence of any peaks of Ca$_3$SiO$_5$ in the raw reaction product [91] in the samples intended to produce Ca$_3$SiO$_5$ (C$_3$S). This is particularly important since in the analysis of the XRD data of the above samples, it always appears the possibility of the existence of trace amounts of hatrurite (Ca$_3$SiO$_5$), something which is not confirmed by Raman spectroscopy.

Relevant to the molten salt synthesis of Ca$_2$SiO$_4$ at ~900°C in 2 hours in this work, it is worth mentioning that the preparation of Ca$_2$SiO$_4$ by solid-state reaction of CaCO$_3$ and SiO$_2$ at 800°C for 10 hours at N$_2$ atmosphere was reported recently [92]. An important result of the reported work on the raw reaction product was the successful preparation of the cement compound β-Ca$_2$SiO$_4$ (β-C$_2$S) (belite) at a temperature of 908°C [88] which is lower than all other previous molten salt synthesis reaction temperatures published so far. This temperature is also lower than the 1000°C (for 12 hours) and 1200°C (for 3 hours) used in a typical solid-state reaction synthesis of belite [93].

A representative Powder XRD measurement of the precipitated raw product of the reaction mixture 1.A is shown in Figure 4 [88]. Assignment of the peaks was after comparison with the XRD data for NaCl and CaO from Wyckoff [94] and for β-Ca$_2$SiO$_4$ from Tsurumi [95] respectively.

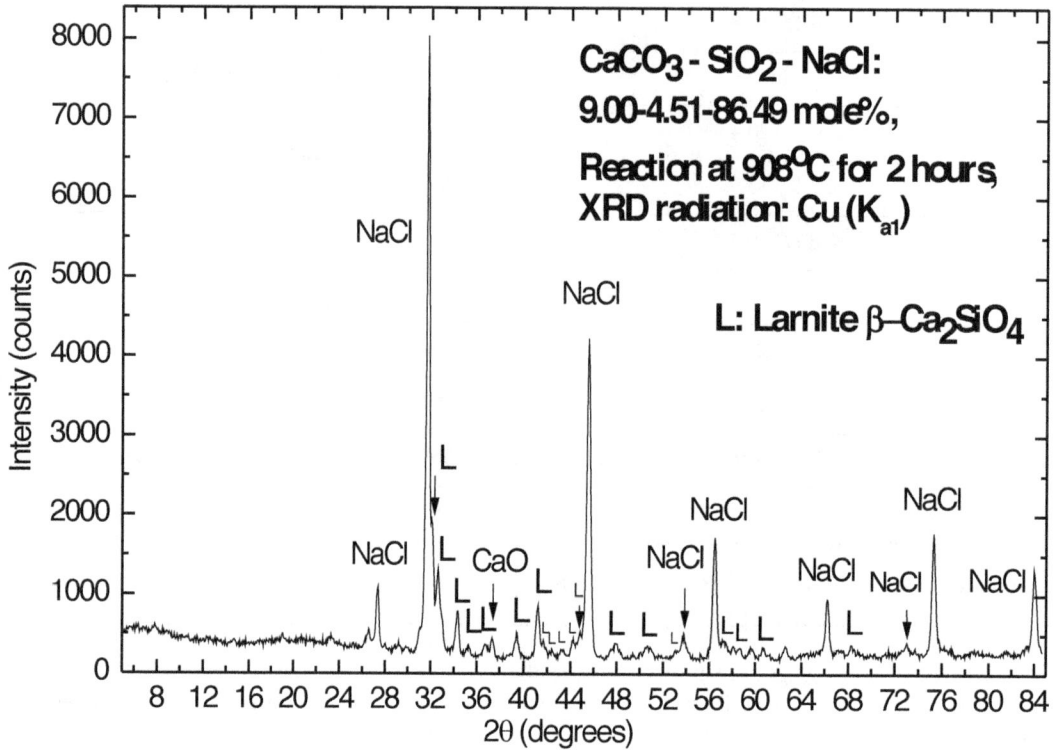

Figure 4. XRD pattern of the raw product of sample 1.A. Republished with permission of Maney Publishing, from Advances in Applied Ceramics, G. Photiadis et al., 110, 3, 2011 [Reference 88].

The Raman spectrum of the raw reaction product of sample 1.A measured in that work and depicted in Figure 5 [88] showed that the major raw reaction product was larnite (β-Ca$_2$SiO$_4$). The Raman-active modes of β-Ca$_2$SiO$_4$ having the highest intensity are the internal modes of vibration of the tetrahedral [SiO$_4$]$^{4-}$ orthosilicate molecular anion. The Raman peaks were attributed to the following normal modes of vibration [96]:

Stretching: v_1 (A$_1$)=860 and 845 cm^{-1}.

Bending: v_2 (E) = 426 and 415 cm^{-1}..

Stretching: v_3 (F$_2$)= 978, 914 and 897 cm^{-1}.

Bending: v_4 (F$_2$)= 557, 539 and 519 cm^{-1}.

In addition, the Raman spectrum showed a pronounced effect of the atmospheric CO$_2$ on the sample of the raw reaction product, leading to the conversion of a certain amount of CaO (lime)

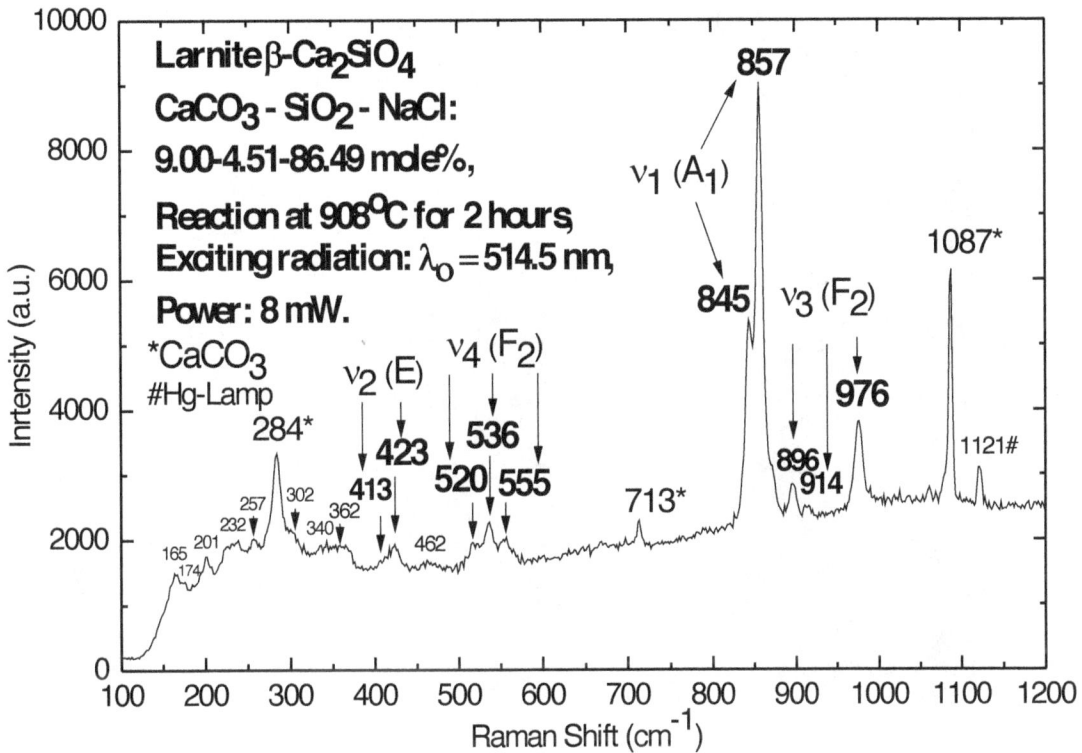

Figure 5. Raman spectrum of raw product of sample 1.A. Republished with permission of Maney Publishing, from Advances in Applied Ceramics, G. Photiadis et al., 110, 3, 2011 [Reference 88].

to $CaCO_3$ (calcite). Similar data measured in this work show the potential of Raman spectroscopy as a fast and reliable probe of the molten salt synthesis of cement compounds.

The SEM microphotographs of the raw reaction product show that the β-Ca_2SiO_4 (β-C_2S) has the shape of milky droplet globules (Figure 6).

The individual cement compounds are mixed oxides. If their component oxides have very low solubility in a molten salt solvent, the same is approximately the case for the mixed oxide cement compound. That means that the product does not dissolve in the molten salt solvent and can be separated by a technique such as molten salt filtration. Simultaneously the very low solubility of the reactants in the molten salt solvent does not impose an important problem because even this very low solubility is adequate to drive the reaction in completion in a few hours, c.g in 3 hours at 900°C as compared to the 12-24 hours at 1400°C needed in solid-state reaction synthesis.

The solubility of CaO in NaCl is approximately 0.001 mole % at 850°C [97]. The solubility of SiO_2 in molten NaCl is estimated to have a value of similar order of magnitude [98]. Thus since the solubilities of both CaO and SiO_2 in molten NaCl are very low and the solubility of Ca_2SiO_4 in fused NaCl is anticipated to have a value of an approximately similar order of magnitude. This fact explains clearly the absence of any calcium silicate chloride in the raw reaction product.

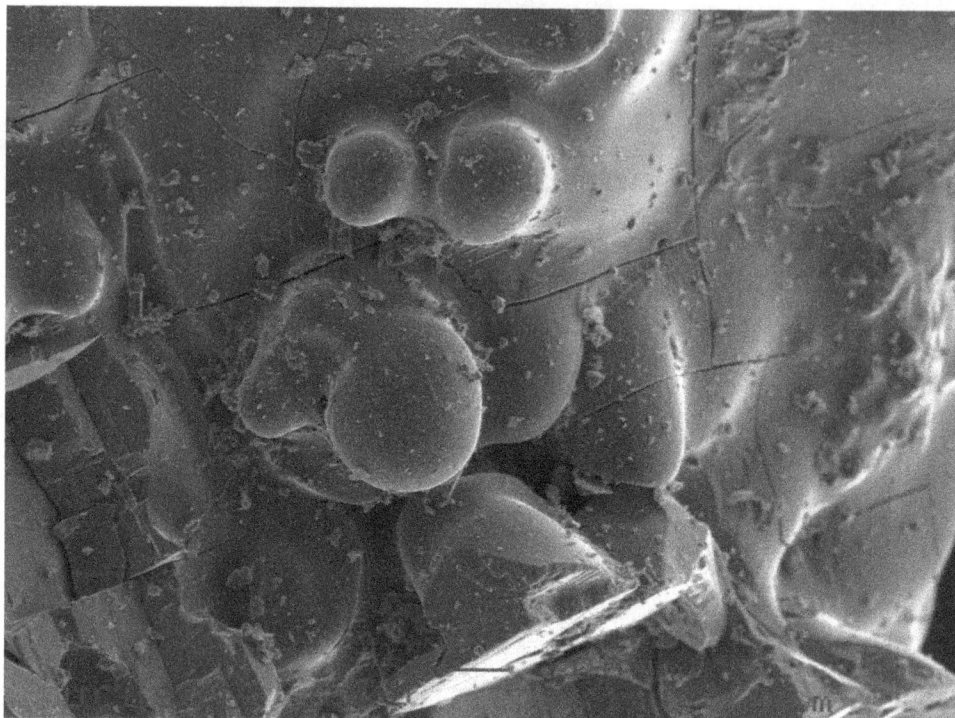

Figure 6. SEM microphotograph of raw product of sample 1.A. The globules of the dicalcium silicate product have size in the order of 50 micrometers.

In order to have a successful synthesis in a molten salt medium it is not a necessary condition to have both reactants absolutely dissolved in the molten salt solvent. In a recent publication on the molten salt synthesis of $MgAl_2O_4$ in molten ACl solvents, SEM pictures of the $MgAl_2O_4$ product show a close resemblance to the SEM pictures of the Al_2O_3 reactant which acts as a template, indicating a 'template' reaction mechanism [73]. Despite the fact that both the CaO and the SiO_2 reactants have a very low solubility in molten ACl (A=Na, K) solvents, their solubility is high enough to yield with high efficiency the Ca_2SiO_4 product in approximately a few hours in a typical experiment at ~900°C. This is a major progress compared to a typical solid-state reaction where tens of hours and reaction temperatures well above 1200°C are necessary in order to produce Ca_2SiO_4. The advantages of molten salt synthesis compared to solid-state reaction synthesis lies in the fact that in the former case the reactants are brought much easier into close contact with each other resulting in higher reaction rates in lower temperatures. It is natural to expect the diffusion of an ion (Ca^{2+} or O^{2-}) to be faster in a liquid medium than in a solid medium. The following reaction mechanism seems to take place. The O^{2-} anions from CaO dissolved in molten NaCl solvent diffuse in the melt and react with SiO_2 forming SiO_4^{2-} molecular anions. In the next stage the Ca^{2+} cations from CaO dissolved in molten NaCl solvent diffuse in the melt and are attached to the SiO_4^{2-} molecular anions to fulfil the electro-neutrality principle, thus forming the Ca_2SiO_4 crystals.

In order to separate the reaction product (β-Ca_2SiO_4) from the molten salt solvent (NaCl) two methods were used:

1. Dissolution of the raw reaction product in water and subsequent filtration of the aqueous solution resulting in the removal of NaCl.

2. High temperature filtration of the raw reaction product, exploiting the fact that at temperatures of ~900°C the β-Ca_2SiO_4 product is solid while the NaCl molten salt solvent is liquid.

The so-called 'wet' method using a solvent like water to wash out the NaCl molten salt solvent is applicable only in the case of β-Ca_2SiO_4 (β-C_2S) samples, because β-Ca_2SiO_4 (β-C_2S) is well known to be much less hygroscopic than Ca_3SiO_5 (C_3S).

Typical X-Ray diffractograms obtained for the water filtered samples are shown in Figure 7 and in Figure 8.

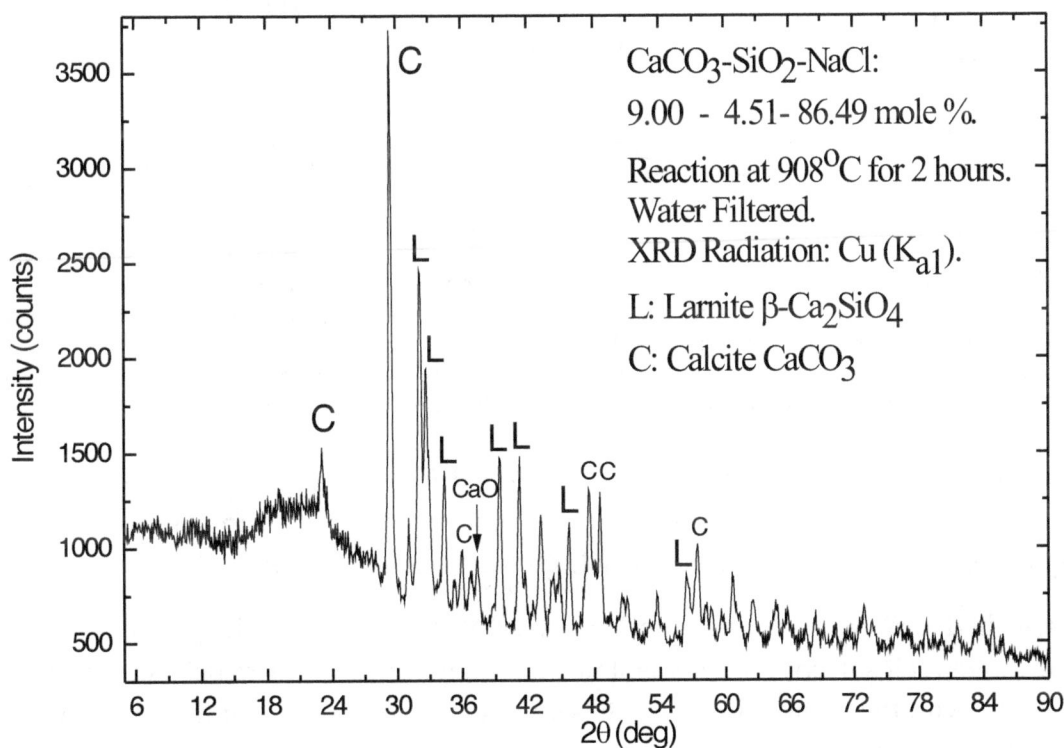

Figure 7. XRD pattern of the water filtered powder product of sample 1.A. The aqueous solution was heated at 80°C in the ultrasonic bath for ~30 minutes and the precipitate having the form of light green grains was dried at 100°C for 3hours and 45 minutes.

It is evident from the diffractogram in Figure 7 that the excess CaO in the raw product sample has reacted with carbon dioxide (CO_2) forming calcite ($CaCO_3$). The presence of peaks of medium to very strong intensity due to calcite ($CaCO_3$) in the XRD pattern in Figure 7 shows that the reaction was not complete but was not far from full completion, taking into account that the reaction time was just 2 hours. In an incomplete reaction the amount of free lime CaO which has not reacted with SiO_2, readily absorbs atmospheric CO_2 and transforms to calcite $CaCO_3$.

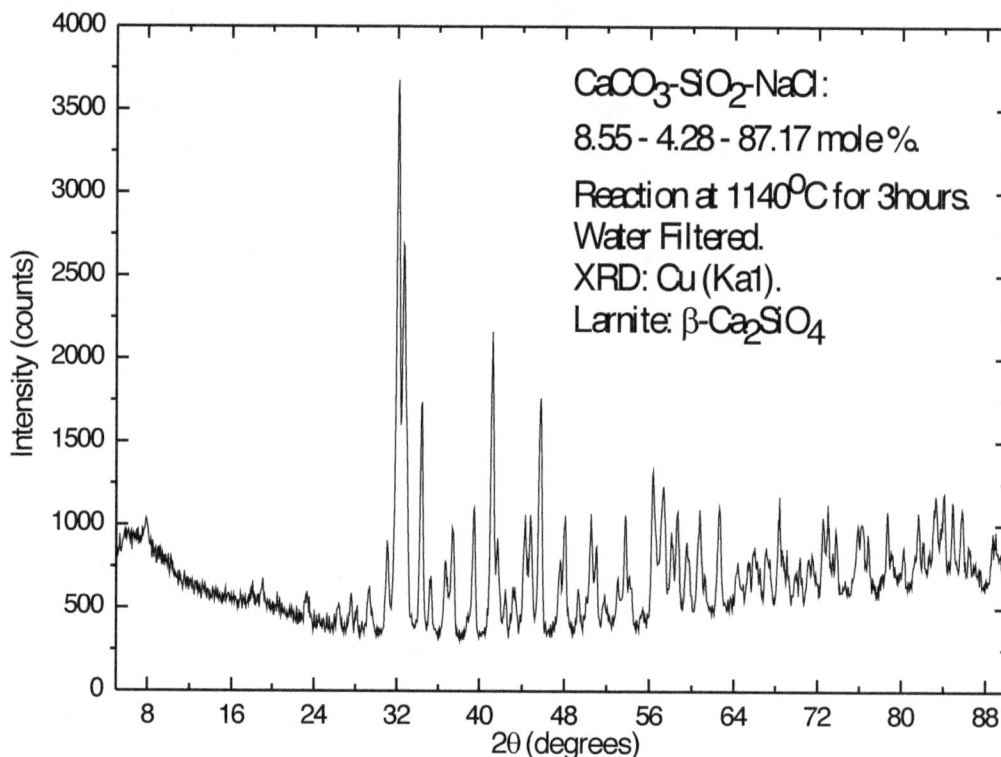

Figure 8. XRD pattern of the water filtered powder product of sample 3. Adding water to the reaction mixture the solution had pH=11 (compared to the pH=14 of water filtered sample 1.A) showing the absence of free lime which would give calcium hydroxide. The aqueous solution was heated at 80°C in the ultrasonic bath for ~30 minutes. The filter paper was glass microfibre FB59407, MF100 (Fisherbrand, Fisher Scientific UK Ltd) made from 100% borosilicate glass, with a retention of 1.6 μm. The precipitate having the form of light green grains was dried at 107°C overnight.

In the XRD pattern of Figure 8 it is evident that the diffractogram is dominated by the peaks of larnite. The absence of peaks due to lime CaO or to calcite $CaCO_3$ confirms that the reaction is complete as is anticipated from the very high reaction temperature.

In the so-called 'dry' method involving "cupelling" the separation of the precipitated product from the molten salt solvent is implemented using Magnesia (MgO) 'cupels' which absorb the molten NaCl with the β-Dicalcium Silicate remaining on the surface of the cupel. It is highly recommended to use 'cupelling' only after the reactions products are known, because cupel may absorb not only the molten salt solvent but also the CaO and SiO_2 powders that have not reacted. In general physical methods of separation including molten salt filtration / centrifugation techniques and high-vacuum sublimation/distillation) are indeed simpler than chemical methods. The X-Ray diffractogram of sample 1.A after cupellation at 850°C for 4 hours is shown in Figure 9. It is evident from comparison of Figure 7 to Figure 9 that cupellation seems to be a not so efficient separation method as is water filtration. The presence of medium to strong intensity peaks of NaCl is obvious in Figure 9, while these peaks are either absent or of very weak intensity in the case of the water filtered sample in Figure 7.

A comparative view of the XRD diffractograms for the raw product, the cupelled and the water filtered sample 1.A. is presented in Figure 10. The differences between the 'processed' product

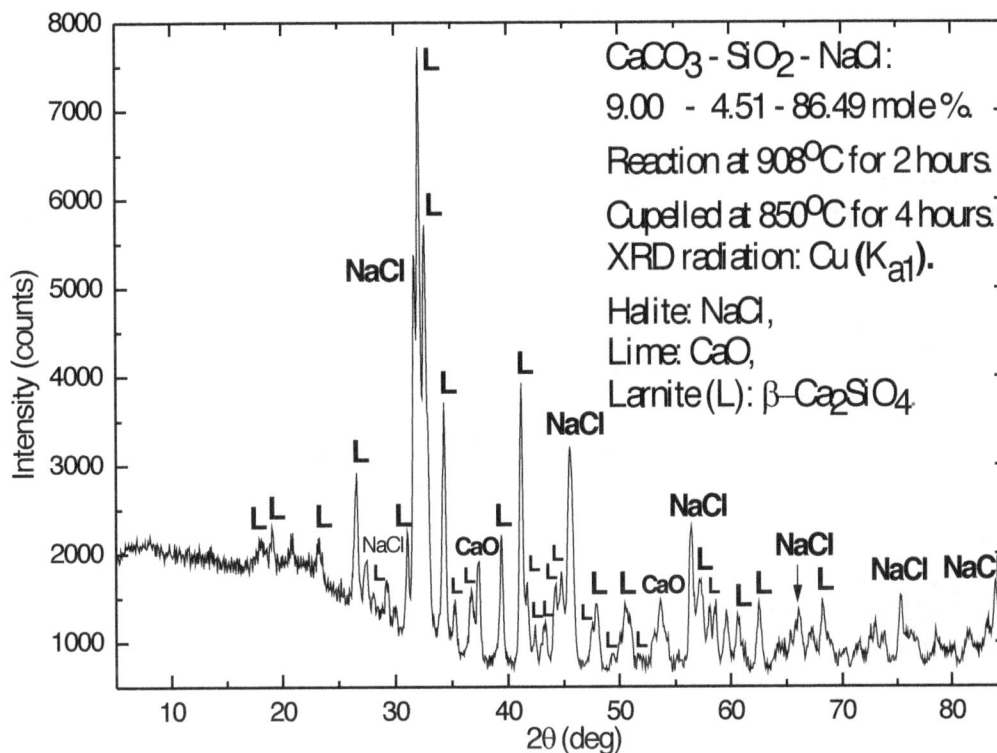

Figure 9. XRD pattern of the cupelled sample 1.A. Cupellation at 850°C for 4 hours.

employing the two separation methods and between them and the raw reaction product are clearly visible in the relative intensities of the peaks of the individual components in Figure 10. Similar XRD patterns of the raw product, the cupelled and the water filtered of other samples provide important information on the influence of the separation technique on the outcome of the reaction.

Preliminary Raman results show that there is some very small amount of γ-Ca$_2$SiO$_4$ present along with the dominant phase of β-Ca$_2$SiO$_4$ (a Raman peak at 814 cm^{-1} shows the presence of trace γ-Ca$_2$SiO$_4$), in the water filtered product.

5.3. The potential reductions in CO$_2$ emissions using the method of molten salt synthesis for the production of cement compounds on an industrial scale

The cement industry is responsible for 5% of the global anthropogenic CO$_2$ emissions [34]. The Ordinary Portland Cement (OPC) consists of 95 wt.% Clinker (3CaO·SiO$_2$, 2CaO·SiO$_2$, 3CaO·Al$_2$O$_3$, 4CaO· Al$_2$O$_3$·Fe$_2$O$_3$) and of 5 wt.% Gypsum (CaSO$_4$·2H$_2$O) [34]. The average intensity of carbon dioxide emissions from the total global cement production was reported to be 222 kg of C/t of cement or 814 kg of CO$_2$/t of cement [34]. The highly energy-intensive process of cement production consists of three major steps: raw material preparation, clinker making in the kiln and cement making [34]. The preparation of raw material involving crushing and grinding the reactants is a process consuming electricity. The clinker kiln uses nearly all of the fuel in a typical cement plant and the production of clinker is the most energy-intensive

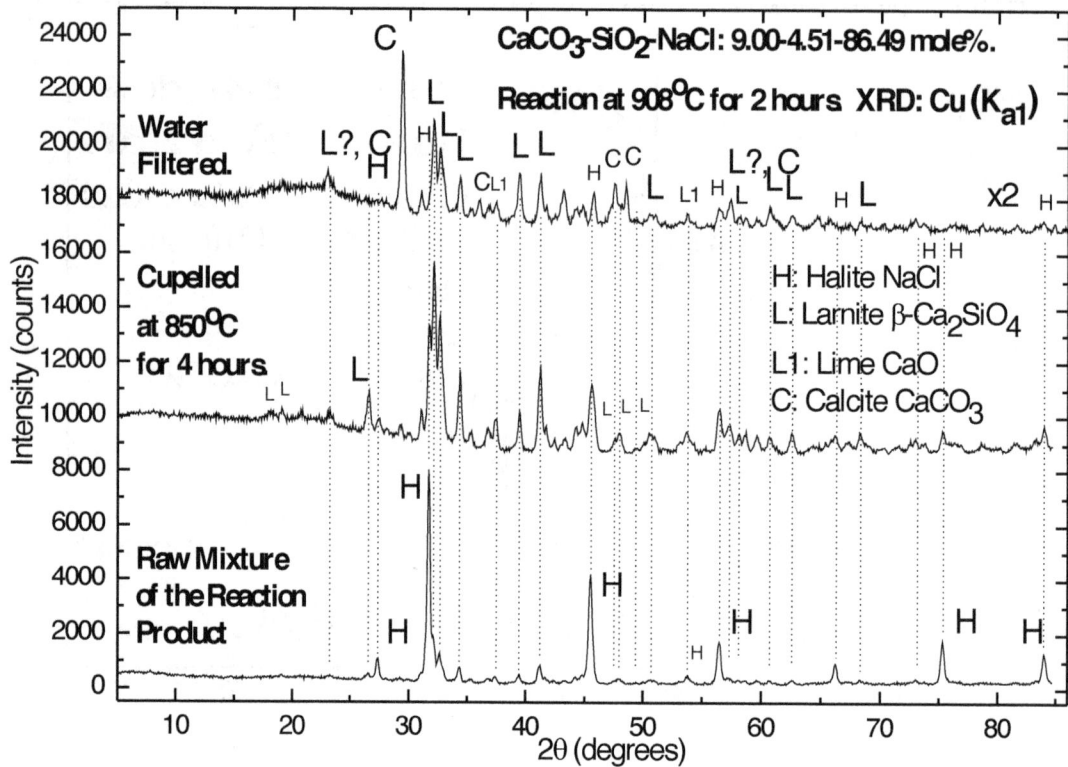

Figure 10. XRD patterns of the raw product, of the cupelled and of the water filtered sample 1.A.

production step, responsible for about 70%–80% of the total energy consumed. The cement making involving finish grinding is a process consuming electricity [34]. The process related CO_2 emissions are due to the decomposition of limestone ($CaCO_3$) to lime (CaO) and to carbon dioxide (CO_2). These process related CO_2 emissions from clinker production amount to about 0.5 kg CO_2/kg of clinker [34] and represent roughly 60% (500/814) of the total emissions. The remaining 40% are the energy related emissions of the cement production process in all of the three major steps of cement manufacture mentioned above.

The suggested low carbon low energy molten salt synthesis of cement compounds has the potential to lead to reductions in both process and energy related carbon dioxide emissions.

Concerning the process related carbon dioxide emissions, the molten salt synthesis method has the potential to use technology enabling the total capture and storage of all carbon dioxide emissions from the calcination of limestone ($CaCO_3$). A precalciner molten salt reactor is a realistic option to achieve the decomposition of limestone ($CaCO_3$) to lime (CaO) and to carbon dioxide(CO_2) in a molten salt solvent and the capture of all carbon dioxide (CO_2). Thus it will be possible to eliminate the process related emissions and to reduce the carbon dioxide emissions of the cement industry by 60%. The elimination of process related emissions within a Molten Salt Synthesis method means a cement industry capable of reducing its impact from 5% to 2% (0.60x0.05 equal to 3% process related emissions gone, 0.40x0.05 equal to 2% energy related emissions left) of the global anthropogenic CO_2 emissions.

The estimation of reduction in carbon dioxide emissions relevant to the energy related emissions is far more complex and has to take into account the production of the individual cement compounds. In the following paragraphs some rough estimates will be attempted by doing some calculations. A rough assumption made is that the temperature, the energy and the carbon dioxide emissions involved in generating the necessary energy to achieve a certain reaction temperature bear all a linear dependence to each other.

The production of β-$2CaO \cdot SiO_2$ with molten salt synthesis at ~900°C (1173 K) represents a reduction of nearly 550°C from the temperature of 1450°C (1723 K) used in a cement kiln. Since β-$2CaO \cdot SiO_2$ is 25-30 wt.% of OPC clinker and the production of clinker is responsible for about 70%–80% of the total energy consumed, and the energy related emissions are the 40% of the total emissions, then the roughly calculated reduction in carbon dioxide emissions can be from 2.23% [(550/1723)x0.25x0.70x0.40] up to 3.06%[(550/1723)x0.30x0.80x0.40] of the emissions of the cement industry or from 0.11% [(550/1723)x0.25x0.70x0.40x0.05] up to 0.15% [(550/1723)x0.30x0.80x0.40x0.05] in the global anthropogenic CO_2 emissions of the cement industry

The production of $12CaO \cdot 7Al_2O_3$ with molten salt synthesis at ~900°C (1173 K) represents a reduction of nearly 550°C from the temperature of 1450°C (1723 K) used in a cement kiln.

The $12CaO \cdot 7Al_2O_3$ compound is an intermediate compound during the synthesis of $3CaO \cdot Al_2O_3$. The $12CaO \cdot 7Al_2O_3$ compound is formed with molten salt synthesis at ~900°C (1173 K) while the synthesis of $3CaO \cdot Al_2O_3$ in the kiln is taking place at ~1250°C (1523 K). The ratio 0.77 of the two temperatures of the reactions of syntheses can be used as a very rough estimate of how the low carbon low energy molten salt synthesis of $12CaO \cdot 7Al_2O_3$ may facilitate the formation of $3CaO \cdot Al_2O_3$ [0.77=(1173/1523)]. Since $12CaO \cdot 7Al_2O_3$ is an intermediate compound during the synthesis of $3CaO \cdot Al_2O_3$ which is 5-12 wt.% of OPC clinker and the production of clinker is responsible for about 70%–80% of the total energy consumed, and the energy related emissions are the 40% of the total emissions, then the roughly calculated reduction in carbon dioxide emissions can be from 0.34% [(550/1723)x0.05x0.70x0.40x0.77] up to 0.94% [(550/1723)x0.12x0.80x0.40x0.77] of the emissions of the cement industry. or from 0.02% [(550/1723)x0.05x0.70x0.40x0.77x0.05] up to 0.05% [(550/1723)x0.12x0.80x0.40x0.77x0.05] in the global anthropogenic CO_2 emissions of the cement industry

Thus the use of molten salt synthesis in the production of β-$2CaO \cdot SiO_2$ and of $12CaO \cdot 7Al_2O_3$ can lead to a roughly estimated reduction of the carbon dioxide emissions of the cement industry from 2.57% (2.23%+0.34%) up to 4% (3.06%+0.94%). The use of molten salt synthesis method for the production of the above two compounds represents a potential reduction of up to 0.2% (0.04x0.05) in the global anthropogenic CO_2 emissions of the cement industry. It is also possible to achieve further progress by attempting the molten salt synthesis of $3CaO \cdot SiO_2$ and of $3CaO \cdot Al_2O_3$ at temperatures above the 1140°C used in the studies reported here, but below the temperature of 1450°C used in the kiln. In a rough calculation a potential molten salt synthesis of $3CaO \cdot SiO_2$ and of $3CaO \cdot Al_2O_3$ at 1250°C could reduce the global anthropogenic CO_2 emissions of the cement industry by up to 0.13% [(200/1723)x0.70x0.80x0.40x0.05] for $3CaO \cdot SiO_2$ and by up to 0.02% [(200/1723)x0.12x0.80x0.40x0.05] for $3CaO \cdot Al_2O_3$.

It is also suggested that the controlled molten salt synthesis of cement compounds with desired particle size distribution can reduce the energy needs (and the associated carbon dioxide emissions) for the finish grinding in the final step of cement making.

The above rough but not far from reality calculations show that the adoption of the molten salt synthesis method in the production of cement compounds offer a realistic alternative option and a complementary to other approach to mitigate the environmental impact of the cement industry. It can lead to a cement industry responsible for less than 1.8% (5%-3%-0.2%) of the global anthropogenic CO_2 emissions which is a figure much better than the current 5%.

6. Conclusion

The compound β-Ca_2SiO_4 (β-$2CaO \cdot SiO_2$) which is an important constituent of Portland cement and a major component of belitic cements was synthesized by the reaction of $CaCO_3$ (calcite) with SiO_2 (α-quartz) in molten NaCl solvent at ~900°C [88, 89]. This is the lowest reaction temperature in molten salt media in which this product has been formed when using $CaCO_3$ and SiO_2 as reactants. The β-Ca_2SiO_4 polymorph has been stabilized at room temperature without the need to use any kind of doping with B_2O_3, Al_2O_3 or sulfates to prevent it's inversion to γ-Ca_2SiO_4. The synthesis of β-Ca_2SiO_4 was also achieved using a slow cooling rate of the reaction mixture, just a few °C/minute (e.g. 3°C/minute) which is much slower than the quenching used in the industrial process of cement manufacture. The effect of two separation methods on removing the molten salt solvent from the reaction product was studied by powder XRD. This is of particular importance since it is necessary to have a β-Dicalcium Silicate product with the minimum amount of NaCl present. Similar very promising results have been reported for the synthesis of $Ca_{12}Al_{14}O_{33}$ ($12CaO \cdot 7Al_2O_3$) in molten NaCl at ~900°C [89, 90]. The Dodecacalcium Heptaaluminate ($12CaO \cdot 7Al_2O_3$) is an intermediate product during the formation of Tricalcium Aluminate ($3CaO \cdot Al_2O_3$) which is a major component of calcium aluminate eco-friendly cements.

The energetically efficient synthesis of β-C_2S and of $C_{12}A_7$ in molten alkali chloride solvents at temperatures as low as 900°C is very promising and paves the way for electric energy powered cement production processes. This is a breakthrough result which can lead to a much more efficient use of energy in the production of cement. It can also pose much less technological challenges for the decomposition of limestone to lime and the efficient capture and storage of the emitted CO_2, thus enabling the use of a molten salt reactor precalciner which can reduce directly the global carbon dioxide emissions of the cement industry from 5% to 2%. The benefits of using a molten salt synthesis process for cement manufacture are that it can reduce both the process and the energy related emissions of carbon dioxide. It can result in a cement industry responsible for less than 1.8% (5%-3%-0.2%) of the global anthropogenic CO_2 emissions, a figure which represents not a simple step change but a significant improvement compared to the 5% of the current cement industry. The strategies to mitigate climate change already in use or planned for use in the future by the cement industry were assessed in the report of the IPCC in 2007 on Mitigation of Climate Change [reference 10, page 467]. In this report it was emphasized that the cement industry is capital intensive and the equipment has a long lifetime, which is a factor limiting the economic potential in the short term. In this respect, an industrial process using the method of the molten salt synthesis of cements has the potential on a medium to longer term to be a realistic novel technology.

Acknowledgements

The work on the molten salt synthesis of cement compounds reported in this chapter has been funded by the Engineering and Physical Sciences Research Council (EPSRC) under grant no. EP/F014449/1 (http://gow.epsrc.ac.uk/ViewGrant.aspx?GrantRef5EP/F014449/1). The author gratefully acknowledges the support of S.J.R. Simons in providing the facilities for the experimental work on the preparation of the samples in the Centre for CO_2 Technology, Department of Chemical Engineering, University College London, London, UK. The author also gratefully acknowledges the support of P. Barnes, J. K. Cockcroft and M. Vickers in the Materials Chemistry Centre, Department of Chemistry, University College, London, UK, for providing the powder XRD facilities used in the study. The author is indebted to cement chemist John Bensted for discussions on cement chemistry and industry. This chapter is dedicated to the memory of my mother Sophia Photiadou who supported my endeavours in science and in particular in Raman Spectroscopy and in Molten Salt Chemistry and Technology. Opinions expressed in this chapter reflect purely the author's view on various issues and do not necessarily represent the official views of the EPSRC and of University College London. The affiliation of the author during the work on molten salt synthesis of cement compounds reported in this chapter was with the Centre for CO_2 Technology, Department of Chemical Engineering, University College London, London, United Kingdom.

Author details

Georgios M. Photiadis[*]

Address all correspondence to: georgios.photiadis@gmail.com

Centre for CO_2 Technology, Department of Chemical Engineering, University College London, London, United Kingdom

*Current Address: Scientist in Raman Spectroscopy and in Molten Salt Chemistry and Technology, Potters Bar, Hertfordshire, England, United Kingdom.

References

[1] Wikipedia: Global Warming. http://en.wikipedia.org/wiki/Global_warming, (accessed 1 October 2014).

[2]   Wikipedia: Svante Arrhenius. http://en.wikipedia.org/wiki/Svante_Arrhenius, (accessed 1 October 2014).

[3]   Arrhenius S. On the Influence of Carbonic Acid in the Air Upon the Temperature of the Ground. Philosophical Magazine, 1896; 41, 237-276. http://www.globalwarmingart.com/images/1/18/Arrhenius.pdf (accessed 1 October 2014).

[4]   Intergovernmental Panel on Climate Change (IPCC). http://www.ipcc.ch/ (accessed 02 October 2014).

[5]   Intergovernmental Panel on Climate Change (IPCC). Organization. http://www.ipcc.ch/organization/organization.shtml (accessed 2 October 2014).

[6]   Solomon S., Qin, D., Manning M., Chen Z., Marquis M., Averyt K.B., Tignor M. and Miller H.L. (eds.). IPCC, 2007: Climate Change 2007: The Physical Science Basis. Contribution of Working Group I to the Fourth Assessment Report of the Intergovernmental Panel on Climate Change. Report of the Intergovernmental Panel on Climate Change. Cambridge University Press, Cambridge, United Kingdom and New York, NY, USA, 996 pp. Front Matter. https://www.ipcc-wg1.unibe.ch/publications/wg1-ar4/ar4-wg1-frontmatter.pdf (accessed 1 October 2014).

[7]   Intergovernmental Panel on Climate Change (IPCC). Working Groups / Task Force. http://www.ipcc.ch/working_groups/working_groups.shtml (accessed 2 October 2014).

[8]   Solomon S., Qin, D., Manning M., Chen Z., Marquis M., Averyt K.B., Tignor M. and Miller H.L., (eds.). IPCC, 2007: Climate Change 2007: The Physical Science Basis. Contribution of Working Group I to the Fourth Assessment Report of the Intergovernmental Panel on Climate Change. Report of the Intergovernmental Panel on Climate Change. Cambridge University Press, Cambridge, United Kingdom and New York, NY, USA, 996 pp. https://www.ipcc-wg1.unibe.ch/publications/wg1-ar4/wg1-ar4.html (accessed 1 October 2014).

[9]   Parry M.L., Canziani O.F., Palutikof J.P., van der Linden P.J.and Hanson C.E., (eds.). IPCC, 2007: Climate Change 2007: Impacts, Adaptation and Vulnerability. Contribution of Working Group II to the Fourth Assessment Report of the Intergovernmental Panel on Climate Change. Cambridge University Press, Cambridge, UK, 976pp. http://www.ipcc.ch/pdf/assessment-report/ar4/wg2/ar4_wg2_full_report.pdf (accessed 2 October 2014).

[10] Metz B., Davidson O.R., Bosch P.R., Dave R., L.A. Meyer L.A., (eds.). IPCC, 2007: Climate Change 2007: Mitigation. Contribution of Working Group III to the Fourth Assessment Report of the Intergovernmental Panel on Climate Change. Cambridge University Press, Cambridge, United Kingdom and New York, NY, USA., 863 pp. http://www.ipcc.ch/pdf/assessment-report/ar4/wg3/ar4_wg3_full_report.pdf (accessed 2 October 2014). See p.467 in Minerals for Cement

[11] Solomon S., Qin, D., Manning M., Chen Z., Marquis M., Averyt K.B., Tignor M. and Miller H.L. (eds.). IPCC, 2007: Climate Change 2007: The Physical Science Basis. Con-

tribution of Working Group I to the Fourth Assessment Report of the Intergovern-mental Panel on Climate Change. Cambridge University Press, Cambridge, United Kingdom and New York, NY, USA, p.98. https://www.ipcc-wg1.unibe.ch/publica-tions/wg1-ar4/ar4-wg1-faqs.pdf   https://www.ipcc-wg1.unibe.ch/publications/wg1-ar4/faq/wg1_faq-1.3.html (accessed 1 October 2014).

[12] Solomon S., Qin, D., Manning M., Chen Z., Marquis M., Averyt K.B., Tignor M. and Miller H.L. (eds.). IPCC, 2007: Climate Change 2007: The Physical Science Basis. Con-tribution of Working Group I to the Fourth Assessment Report of the Intergovern-mental Panel on Climate Change. Cambridge University Press, Cambridge, United Kingdom and New York, NY, USA, p.100. https://www.ipcc-wg1.unibe.ch/publica-tions/wg1-ar4/ar4-wg1-faqs.pdf   https://www.ipcc-wg1.unibe.ch/publications/wg1-ar4/faq/wg1_faq-2.1.html (accessed 1 October 2014).

[13] Solomon, S., Qin D., Manning M., Alley R.B., Berntsen T., Bindoff N.L., Chen Z., Chidthaisong A., Gregory J.M., Hegerl G.C., Heimann M., Hewitson B., Hoskins B.J., Joos F., Jouzel J., Kattsov V., Lohmann U., Matsuno T., Molina M., Nicholls N., Over-peck J., Raga G., Ramaswamy V., Ren J., Rusticucci M., Somerville R., Stocker T.F., Whetton P., Wood R.A. and Wratt D. 2007: Technical Summary, p.25. In: Solomon S., Qin, D., Manning M., Chen Z., Marquis M., Averyt K.B., Tignor M. and Miller H.L. (eds.). Climate Change 2007: The Physical Science Basis. Contribution of Working Group I to the Fourth Assessment Report of the Intergovernmental Panel on Climate Change. Cambridge University Press, Cambridge, United Kingdom and New York, NY, USA. https://www.ipcc-wg1.unibe.ch/publications/wg1-ar4/ar4-wg1-ts.pdf (ac-cessed 1 October 2014).

[14] Stocker, T.F., Qin D., Plattner G.-K., Tignor M., Allen S.K., Boschung J., Nauels A., Xia Y., Bex V. and Midgley P.M., Editors. IPCC, 2013: Summary for Policymakers. In: Climate Change 2013: The Physical Science Basis. Contribution of Working Group I to the Fifth Assessment Report of the Intergovernmental Panel on Climate Change. Cambridge University Press, Cambridge, United Kingdom and New York, NY, USA, pp. 1–30. doi:10.1017/CBO9781107415324.004 http://www.climatechange2013.org/images/report/WG1AR5_SPM_FINAL.pdf (accessed 4 October 2014).

[15] Wikipedia: Keeling Curve. http://en.wikipedia.org/wiki/Keeling_Curve (accessed 7 October 2014).

[16] BBC News: 50 years on: The Keeling Curve legacy. By Helen Briggs, Science reporter, BBC News. Page last updated at 20:13 GMT, Sunday, 2 December 2007. http://news.bbc.co.uk/1/hi/sci/tech/7120770.stm (assessed 7 October 2014).

[17] Wikipedia: Kyoto Protocol. http://en.wikipedia.org/wiki/Kyoto_Protocol (accessed 7 October 2014).

[18] Dr. James E. Hansen: Climate Science Awareness and Solutions program. Earth Insti-tute, Columbia University, New York, USA. http://www.columbia.edu/~jeh1/ (as-sessed 7 October 2014).

[19] James E. Hansen: Dangerous Anthropogenic Interference-A Discussion of Humanity's Faustian Climate Bargain and the Payments Coming Due. Presentation on October 26, 2004, in the Distinguished Public Lecture Series at the Department of Physics and Astronomy, University of Iowa. http://www.columbia.edu/~jeh1/2004/dai_complete_20041026.pdf (assessed 7 October 2014.

[20] Hansen J, Sato M., Ruedy R. Lo K., Lea D.W., Medina-Elizade M. Global temperature change. Proceedings of the National Academy of Sciences, 2006; 103(39), 14288-14293. www.pnas.org_cgi_doi_10.1073_pnas.0606291103.   doi:10.1073/pnas.0606291103 http://pubs.giss.nasa.gov/docs/2006/2006_Hansen_etal_1.pdf (assessed 7 October 2014).

[21] Hansen J., Ruedy R., Sato M. Lo K. Global Surface Temperature Change. Reviews of Geophysics, 2010; 48(4), RG4004. DOI: 10.1029/2010RG000345. http://data.giss.nasa.gov/gistemp/paper/gistemp2010_draft0803.pdf (assessed 7 October 2014).

[22] Hansen J., Kharecha P., Sato M., Masson-Delmotte V., Ackerman F., Beerling D.J., Hearty P.J., Hoegh-Guldberg O., Hsu S.-L., Parmesan C., Rockstrom J., Rohling E.J., Sachs J., Smith P., Steffen K., Van Susteren L., von Schuckmann K., C. Zachos J.C. Assessing "Dangerous Climate Change": Required Reduction of Carbon Emissions to Protect Young People, Future Generations and Nature. PLoS ONE, 2013; 8(12), e81648. doi:10.1371/journal.pone.0081648. http://www.plosone.org/article/fetchObject.action?uri=info%3Adoi%2F10.1371%2Fjournal.pone.0081648&representation=PDF (assessed 7 October 2014).

[23] Stocker, T.F., Qin D., Plattner G.-K., Tignor M., Allen S.K, Boschung J., Nauels A., Xia Y., Bex V. and Midgley P.M., Editors. IPCC, 2013: Climate Change 2013: The Physical Science Basis. Contribution of Working Group I to the Fifth Assessment Report of the Intergovernmental Panel on Climate Change. Cambridge University Press, Cambridge, United Kingdom and New York, NY, USA, 1535 pp. http://www.climatechange2013.org/images/report/WG1AR5_ALL_FINAL.pdf (accessed 7 October 2014).

[24] Taylor H. F.W. Cement Chemistry. London: Thomas Telford Publishing; 1997, 2nd Edition, 480 pp.

[25] Hewlett, P. C., editor. Lea's Chemistry of Cement and Concrete. Oxford: Elsevier, Butterworth-Heinemann; 1988, 4th Edition, 1092 pp.

[26] Hansen T.C., Radjy F., Sellevold E.J. Cement Paste and Concrete. Annual Review of Materials Science, 1973; 3, 233-268. DOI: 10.1146/annurev.ms.03.080173.001313.

[27] Portland Cement Association. PCA. America's Cement Manufacturers: How Cement is Made. http://www.cement.org/cement-concrete-basics/how-cement-is-made (accessed 11 October 2014).

[28] U.S. Department of the Interior. U.S. Geological Survey. Minerals Information: Cement Statistics and Information. http://minerals.usgs.gov/minerals/pubs/commodity/

cement/index.html Mineral Commodity Summaries, February 2014. http://minerals.usgs.gov/minerals/pubs/commodity/cement/mcs-2014-cemen.pdf (accessed 11 October 2014).

[29] Galvez-Martos J.-L. Building a 'green construction' powerhouse for the world. Energeia, Spring/Summer 2013; 11-13. University of Aberdeen. http://www.abdn.ac.uk/aie/documents/Energeia_ISSUE_3.pdf (accessed 11 October 2014).

[30] World Population: Past, Present, and Future, worldometers. http://www.worldometers.info/world-population/ (accessed 10 June 2014).

[31] van Oss, H.G. Background Facts and Issues Concerning Cement and Cement Data. Open-File Report 2005-1152, U.S. Department of the Interior, U.S. Geological Survey, USGS, science for a changing world, Published 2005, Online only, Version 1.2 (posted February 9, 2006), i-viii and 1-80 pages. http://pubs.usgs.gov/of/2005/1152/2005-1152.pdf (accessed 11 October 2014).

[32] Pratt P.L., Jennings H.M. The Microchemistry and Microstructure of Portland Cement. Annual Review of Materials Science 1981; 11, 123-149. DOI: 10.1146/annurev.ms.11.080181.001011.

[33] Composition of cement. http://www.engr.psu.edu/ce/courses/ce584/concrete/library/construction/curing/Composition%20of%20cement.htm (accessed 11 October 2014).

[34] Worrell E., Price L., Martin N., Hendriks C., Meida L.O. Carbon Dioxide Emissions from the Global Cement Industry. Annual Review of Energy and the Environment 2001; 26, 303-329. DOI: 10.1146/annurev.energy.26.1.303.

[35] Humphreys K., Mahasenan M. Towards a sustainable cement industry. Sub-study 8: climate change. March 2002. An Independent Study Commissioned by: World business Council for Sustainable Development. http://www.wbcsdcement.org/pdf/final_report8.pdf, (accessed 11 October 2014).

[36] Portland Cement Association. PCA. America's Cement Manufacturers: Cement Sustainability Manufacturing Program. Cement Manufacturing Transparency Reporting. PCA Voluntary Code of Conduct. http://www.cement.org/for-concrete-books-learning/cement-manufacturing/cement-sustainability-manufacturing-program (accessed 11 October 2014).

[37] Kowalski M., Spencer P.J., Neuschütz D. Phase Diagrams, Chapter 3, pp. 21-214. $CaO\text{-}SiO_2$, Fig. 3.70, p.63. $CaO\text{-}Al_2O_3$, Fig. 3.19, p.39. $CaO\text{-}SiO_2\text{-}Al_2O_3$, p.105. $CaO\text{-}SiO_2\text{-}Al_2O_3\text{-}FeO$ and $CaO\text{-}SiO_2\text{-}Al_2O_3\text{-}Fe_2O_3$, pp.154-156. In Slag Atlas, 2nd Edition, Verlag Stahleisen GmbH, Düsseldorf, Germany, on behalf of the European Communities. Edited by Verein Deutscher Eisenhüttenleute; 1995. ISBN 3-514-00457-9. http://mme.iitm.ac.in/shukla/Schlackenatlas.pdf (accessed 24 September 2014).

[38] Physical Constants of Inorganic Compounds, 4-43 to 4-101, CRC Handbook of Chemistry and Physics, 95th Edition, Internet Version 2014-2015, William M. Haynes, Edi-

tor-in-Chief, Thomas J. Bruno, Associate Editor, David R. Lide, Editor, Internet Edition, CRC Press. http://www.hbcpnetbase.com/ (accessed 27 August 2014).

[39] Handbook of Mineralogy, Hatrurite, Ca_3SiO_5. Mineral Data Publishing; 2001, version 1.2. http://www.handbookofmineralogy.com/pdfs/hatrurite.pdf. (accessed 27 August 2014).

[40] Handbook of Mineralogy, Larnite, β-Ca_2SiO_4. 2001 Mineral Data Publishing; 2001, version 1.2. http://rruff.info/doclib/hom/larnite.pdf. (accessed 27 August 2014).

[41] Handbook of Mineralogy, Mayenite, $Ca_{12}Al_{14}O_{33}$. Mineral Data Publishing; 2001-2005, version 1. http://rruff.info/doclib/hom/mayenite.pdf. (accessed 27 August 2014).

[42] Telschow S., Frandsen F., Theisen K. Dam-Johansen K. Cement Formation-A Success Story in a Black Box: High Temperature Phase Formation of Portland Cement Clinker. Industrial and Engineering Chemistry Research 2012; 51(34), 10983-11004. DOI: 10.1021/ie300674j.

[43] Li X., Shen X., Tang M., Li X. Stability of Tricalcium Silicate and Other Primary Phases in Portland Cement Clinker. Industrial and Engineering Chemistry Research 2014; 53(5), 1954-1964. DOI: 10.1021/ie4034076.

[44] Hao Y.-J., Tanaka T. Role of the contact points between particles on the reactivity of solids. The Canadian Journal of Chemical Engineering 1988; 66(5), 761-766. doi: 10.1002/cjce.5450660509.

[45] Khawam A., Flanagan D.R. Solid-State Kinetic Models: Basics and Mathematical Fundamentals. The Journal of Physical Chemistry B 2006, 110(35), 17315-17328. DOI: 10.1021/jp062746a.

[46] Hild, K., Trömel, G. Die Reaktion von Calciumoxyd und Kieselsäure im festen Zustand. Reaction of Calcium Oxide and Silicic Acid in the Solid State. Zeitschrift für anorganische und allgemeine Chemie 1933; 215(3-4), 333–344. doi: 10.1002/zaac. 19332150313

[47] De Keyser W. L. La Synthèse Thermique des Silicates de Calcium. Thermal Synthesis of Calcium Silicates. Bulletin des Sociétés Chimiques Belges 1953; 62(3-4), 235-252. doi: 10.1002/bscb.19530620307.

[48] Fierens P., Picquet P. Kinetic Studies of the Thermal Synthesis of Calcium Silicates Above 1400°C: I, Dynamic Thermal Synthesis of Ca_2SiO_4. Journal of the American Ceramic Society 1975; 58(1-2), 50–51. doi: 10.1111/j.1151-2916.1975.tb18982.x.

[49] Fierens P., Picquet P. (1975), Kinetic Studies of the Thermal Synthesis of Calcium Silicates Above 1400°C: II, Quantitative Kinetics of the Formation of Ca_2SiO_4 in the Presence of a Liquid Phase. Journal of the American Ceramic Society 1975; 58(1-2), 52–54. doi: 10.1111/j.1151-2916.1975.tb18983.x.

[50] Shibata S., Kishi K., Asaga K., Daimon M., Shrestha P.R. Preparation and hydration

of β-C$_2$S without stabilizer. Cement and Concrete Research 1984; 14(3), 323-328. DOI: 10.1016/0008-8846(84)90049-8.

[51] Weisweiler W., Osen E., Eck J., Höfer H. Kinetic studies in the CaO-SiO$_2$-System Part I Mechanism and kinetic data of the reactions between CaO-and SiO$_2$-powder compacts. Cement and Concrete Research 1986; 16(3), 283-295. DOI: 10.1016/0008-8846(86)90103-1.

[52] Wesselsky A., Jensen O.M. Synthesis of pure Portland cement phases. Cement and Concrete Research 2009; 39(11), 973-980. DOI: 10.1016/j.cemconres.2009.07.013.

[53] Donaldson C.H., Williams R.J., Lofgren G. A Sample Holding Technique for Study of Crystal Growth in Silicate Melts. American Mineralogist 1975; 60, 324-326. http://rruff.info/doclib/am/vol60/AM60_324.pdf (accessed 12 October 2014).

[54] Maries A., Rogers P.S. Controlled growth of crystalline silicate fibres. Nature 1975; 256, 401 – 402. doi:10.1038/256401a0.

[55] Maries A., Rogers P.S. Continuous unidirectional crystallization of fibrous metasilicates from melts. Journal of Materials Science 1978; 13(10), 2119–2130.

[56] Campbell M., Maries A., Rogers P.S. The synthesis of fluor-analogues of natural asbestos by unidirectional crystallisation. Nature 1979; 281, 129 – 131. doi: 10.1038/281129a0.

[57] Weston R.M., Rogers P.S. The growth of calcium metasilicate polymorphs from supercooled melts and glasses. Mineralogical Magazine 1978; 42, 325-335. DOI: 10.1180/minmag.1978.042.323.02. http://rruff.info/doclib/MinMag/Volume_42/42-323-325.pdf. (accessed 12 October 2014).

[58] Huang X.-H., Chang J. Low-temperature synthesis of nanocrystalline β-dicalcium silicate with high specific surface area. Journal of Nanoparticle Research 2007; 9(6), 1195–1200.

[59] Fujimori H., Yahata D., Yamaguchi N., Ikeda D., Ioku K., Goto S. Synthesis of calcium silicate by a chelate gel route with aqueous solution of citric acid. Journal of the Ceramic Society of Japan 2001; 109(5), 391-395.

[60] Saravanapavan P., Hench L.L. Mesoporous calcium silicate glasses. I. Synthesis. Journal of Non-Crystalline Solids 2003; 318(1-2), Pages 1-13. DOI: 10.1016/S0022-3093(02)01864-1.

[61] Saravanapavan P., Hench L.L. Mesoporous calcium silicate glasses. II. Textural characterisation. Journal of Non-Crystalline Solids 2003; 318(1-2), 14-26. DOI: 10.1016/S0022-3093(02)01882-3.

[62] Stephan D., Wilhelm P. Synthesis of Pure Cementitious Phases by Sol-Gel Process as Precursor. Z. Anorg. Allg. Chem. 2004; 630(10), 1477-1483. DOI: 10.1002/zaac.200400090.

[63] Chrysafi R., Perraki Th., Kakali G. Sol-gel preparation of 2CaO SiO$_2$. Journal of the European Ceramic Society 2007; 27(2-3), 1707-1710. DOI: 10.1016/j.jeurceramsoc. 2006.05.004.

[64] Wu J., Zhu Y.-J., Cheng G.-F., Huang Y.-H. Microwave-assisted preparation of Ca$_6$Si$_6$O$_{17}$(OH)$_2$ and β-CaSiO$_3$ nanobelts. Materials Research Bulletin 2010; 45(4), 509-512.

[65] RUSNANO. GLOSSARYof NANOtechnology and related TERMS: Pechini method. http://eng.thesaurus.rusnano.com/wiki/article2075 (accessed 12October 2014).

[66] Hong S.H. and Young J.F. Hydration Kinetics and Phase Stability of Dicalcium Silicate Synthesized by the Pechini Process. Journal of the American Ceramic Society 1999; 82(7), 1681–1686. DOI: 10.1111/j.1151-2916.1999.tb01986.x.

[67] Romano J.S., Marcato P.D., Rodrigues F.A. Synthesis and characterization of manganese oxide-doped dicalcium silicates obtained from rice hull ash. Powder Technology 2007; 178(1), 5-9.

[68] Lee S.-J., Kriven W. M. Synthesis and hydration study of Portland cement components prepared by the organic steric entrapment method. Materials and Structures 2005; 38(1), 87-92. DOI: 10.1007/BF02480579.

[69] Tremillion B. Reactions in Solution, An Applied Analytical Approach (Translation by D. Inman). John Wiley & Sons Ltd.: Chichester; 1997. pp.301-400.

[70] Al-Raihani H., Durand B. Chassagneux F., Inman D. A novel preparation of calcia fully stabilised zirconia from molten alkali-metal nitrate. Journal of Materials Chemistry 1996; 6, 495-500.

[71] Volkovich V.A., Griffiths T.R., Fray D.J., Fields M. Wilson P.D. Oxidation of UO$_2$ in molten alkali-metal carbonate mixtures: formation of uranates and diuranates. Journal of the Chemical Society, Faraday Transactions 1996; 92(24), 5059-5065.

[72] Volkovich V.A., Griffiths T.R., Fray D.J., Fields M. Increased oxidation of UO$_2$ in molten alkali-metal carbonate based mixtures by increasing oxygen solubility and by controlled generation of superoxide ions, and evidence for a new sodium uranate. Journal of the Chemical Society, Faraday Transactions 1997; 93(21), 3819-3826.

[73] Zhang S., Jayaseelan D.D., Bhattacharya G., Lee W.E. Molten Salt Synthesis of Magnesium Aluminate (MgAl$_2$O$_4$) Spinel Powder. Journal of the American Ceramic Society 2006, 89(5), 1724–1726.

[74] Kanatzidis M.G., Park Y., Polychalcogenide synthesis in molten salts. Novel one-dimensional compounds in the potassium-copper-sulfur system containing exclusively S$_{42}$-ligands. Journal of the American Chemical Society 1989, 111(10), 3767-3769.

[75] Palchik O., Marking G.M., M. G. Kanatzidis M.G. Exploratory Synthesis in Molten Salts: Role of Flux Basicity in the Stabilization of the Complex Thiogermanates Cs$_4$Pb$_4$Ge$_5$S$_{16}$, K$_2$PbGe$_2$S$_6$, and K$_4$Sn$_3$Ge$_3$S$_{14}$. Inorganic Chemistry (Communication) 2005; 44(12), 4151-4153.

[76] Xu C.-Y., Zhen L., Yang R., and Wang Z.L. Synthesis of Single-Crystalline Niobate Nanorods via Ion-Exchange Based on Molten-Salt Reaction. Journal of the American Chemical Society (Communication) 2007; 129(50), 15444-15445.

[77] Rørvik P.M., Lyngdal T., Sæterli R., van Helvoort A.T.J., Holmestad R., Grande T, Einarsrud M.-A. Influence of Volatile Chlorides on the Molten Salt Synthesis of Ternary Oxide Nanorods and Nanoparticles. Inorganic Chemistry 2008; 47(8), 3173-3181.

[78] Brixner L.H., Babcock K. Inorganic single crystals from reactions in fused salts. Materials Research Bulletin 1968; 3 (10), 817-824.

[79] Hermoneit B., Ziemer B. Silikattechnik 1978; 29, 366.

[80] Hermoneit B., Ziemer B. Abstracts of the 11th IUCr Conference: Warsaw; 1978, 211.

[81] Ziemer B., Hermoneit B., Ladwig G. Silikattechnik 1980; 31, 139.

[82] Sakamoto C., Fujii S., Sugie Y., Ohtani N. Rep. Himeji Inst. Tech. (Japan) 1987; 40A, 124.

[83] Sakamoto C., Fujii S., Sugie Y., Tanaka M., Ohtani N. The Crystal-Growth of Monticellite and Akermanite Using Alkali Chlorides. Yogyo-Kyokai-Shi 1987; 95 (7), 749-752.

[84] von Lampe F. Investigations on Belites and Dicalcium Silicates. I. Investigations of the Oxidation Number and the Crystallographic Position of Mn-Ions in the Crystal Lattice of Mn-Doped Ca_2SiO_4. Zeitschrift für anorganische und allgemeine Chemie 1985, 524, 168-176.

[85] Hermoneit B., Ziemer B. Growth of Ca_2SiO_4 crystals from flux. Acta Crystallographica A 1978; 34, S211.

[86] Hermoneit B., Ziemer B., Malewski G. Single crystal growth and some properties of the new compound $Ca_3Si_2O_7 \cdot 1/3CaCl_2$. Journal of Crystal Growth 1981; 52(2), 660-664.

[87] Hermoneit B., Ziemer B. Crystal growth of calcium silicates-A historical review. Journal of Crystal Growth 1982; 59(3), 567-571.

[88] Photiadis G., Maries A., Tyrer M., Inman D., Bensted J., Simons S., Barnes P. Low Energy Synthesis of Cement Compounds in Molten Salt. Advances in Applied Ceramics 2011; 110(3), 137-141. Special Issue of Advances in Applied Ceramics for Cement and Concrete Science Conference 2009, Leeds 2009.

[89] Photiadis G.M., Simons S.J.R., Bensted J., Inman D., Tyrer M., Maries A. Low Energy Molten Salt Synthesis of Cement Compounds. Abstracts of the EUCHEM Conference on Molten Salts and Ionic Liquids 2010, March 14-19, 2010 in Bamberg, Germany, p. 71, Editors: Peter Wasserscheid, Ken Seddon, Marco Haumann and Dana Demtröder.

[90] Photiadis G.M., Simons S.J.R., Bensted J., Inman D., Tyrer M., Maries A. Low Energy Synthesis of the Cement Compound 12CaO 7Al$_2$O$_3$ in Molten Chloride Solvents. Abstracts of the 30th Cement and Concrete Science Conference, University of Birmingham, Birmingham, UK, 13-15th September 2010.

[91] Tarrida M., Madon M., Le Rolland B., Colombet P. An In-Situ Raman Spectroscopy Study of the Hydration of Tricalcium Silicate. Advanced Cement Based Materials, 1995, 2(1), 15-20.

[92] Wang M., Lee C.-G., Ryu C.-K. CO$_2$ sorption and desorption efficiency of Ca$_2$SiO$_4$. International Journal of Hydrogen Energy, 2008, 33(21), 6368-6372.

[93] De la Torre A. G., Aranda M. A. G. Accuracy in Rietveld quantitative phase analysis of Portland cements. Journal of Applied Crystallography 2003, 36, 1169-1176.

[94] Wyckoff R.W.G. Crystal Structures 1, 85-237. Second edition, Interscience Publishers: New York; 1963.

[95] Crystal structure and hydration of belite. Ceramic Transactions, 1994, 40, 19-25. T. Tsurumi, Y. Hirano, H. Kato, T. Kamiya and M. Daimon.

[96] Handke M., Urban M. IR and Raman Spectra of Alkaline Earth Metals Orthosilicates. Journal Molecular Structure 1982, 79, 353-356.

[97] Boghosian S., Godoe A.A., Mediaas H., Ravlo W., Ostvold T. Oxide Complexes in Alkali-Alkaline-Earth Chloride Melts. Acta Chemica Scandinavica 1991; 45, 145-157.

[98] Ishitsuka T., Nose K. Solubility study on protective oxide films in molten chlorides created by refuse incineration environment. Materials and Corrosion-Werkstoffe und Korrosion 2000; 51(3), 177-181.

6

Investigating the Relative Roles of the Degradation of Land and Global Warming in Amazonia

Sergio H. Franchito, J. P. R. Fernandez and
David Pareja

1. Introduction

Large-scale removal of the tropical rain forest will have significant negative effects on regional water and energy balance, climate and global bio-geochemical cycles. Numerical experiments using General Circulation Models (GCMs) [1, 2, 3 and many others], using statistical-dynamical simple climate models (SDMs) [4, 5, 6] and field observations) [7] have shown that the large-scale deforestation in Amazonia may indeed influence regional climate. Reduction in evapotranspiration and precipitation and an increase in the surface temperature in the tropical region occur when the forest is replaced by pasture.

Projections of future climate given in IPCC AR5 (2013) (to be published) indicated that climate change due to anthropogenic human activities is affecting adversely the ecosystems. Many model studies showed that the global warming may affect the biomes distribution over South America, where significant portions of rain forest may be replaced by nonforested areas [8, 9, 10, 11]. These studies suggest that due to increase of greenhouse gases concentration the process of savannization of the tropical forest can be accelerated. This indicates that the future distribution of biomes in the tropical region depends on the combination of the effects of the degradation of land surface and climate changes due to global warming. Some studies have been made to investigate the relative roles of future changes in greenhouse gases compared with future changes in land cover. [12] and [13] compared the climate change simulated under a 2050 SRES B2 greenhouse gases scenario to the one under a 2050 SRES B2 land cover change scenario. It was noted that the relative impact of vegetation change compared to greenhouse gas concentration increase was of the order of 10%, and could reach 30% over limited areas of tropical region. The same methodology was applied for the SRES A2 and B1 scenario over the 2000 to 2100 period [14]. It was also found that although there was no significant effect at the global scale, a large effect at the regional scale may occur, such as a warming of 2°C by 2100 over the Amazon for the A2 land cover change scenario. Recently, studies using SDMs showed that the percentage of the warming due to deforestation relative to the warming when greenhouse gas concentration increase was included together was around 60% in the tropical region [5, 6]. These results suggest that the climate change due to land cover changes may be important relative to the change due to greenhouse gases at the regional level, where intense land cover change occurs. Globally, however, the impact of greenhouse gas concentrations seems to dominate over the impact of land cover change.

Although GCMs and SDMs can provide useful information regarding the response of the global circulation to large-scale forcing, due to their coarse resolution the mesoscale forcing, such as complex topography, vegetation cover, lakes, etc, are not well represented. In this sense Regional Climate Models (RCMs) may be more adequate. RCMs have therefore been developed to downscale larger scale simulations and to provide predictions for specific regions [15, 16, 17, 18].

In this paper the relative roles of the land surface degradation in Amazonia and global warming are investigated using a RCM. The purpose is to inquire how is the effect on the regional climate and aridity due to deforestation and when the increase of concentration of greenhouse gases is also taken into account together. The model to be used is The Abdus Salam International Centre for Theoretical Physics Regional Climate Model v. 4 (ICTP/RegCM4) [19]. In order to take into account the effect of global warming the model will be run using a methodology for generating surrogate climate-change scenarios with a regional climate model [20]. The distribution of aridity is determined using the radiative dryness index of Budyko (AI_B) [21] and the UNEP aridity index (AI_U) [22]. A brief description of the RCM, the methodology employed and the experiments design are given in section 2; the model simulations are presented in section 3 and section 4 contains the summary and conclusions.

2. Regional climate change model

The model ICTP RegCM4 [19] is the version 4 of the regional climate model (RegCM) originally developed at the National Center for Atmospheric Research (NCAR) [15, 16]. The dynamic component of the model is based on the NCAR-Pennsylvania State University meso-escale model (MM5) [23]. For application in climate studies, a number of physical parameterizations were incorporated in the model. More details about the model and physical configurations for South America is given in [19]. In the present study modified parameters of BATS land-surface model for vegetation type 6 (tropical rain forest) are used to reduce the rainfall dry bias over tropical South America, as reported in earlier RegCM versions [24].

The model domain covers the entire South America (Fig. 1), following the CORDEX, an international effort to downscale climate projections over the world using RCMs [25]. The model domain is centered at 22S, 59W, and comprises 202EWx192NS grid points, with a horizontal grid spacing of 50 km over a rotated Mercator projection. Ten-yr simulations were performed (after discarding a 1 yr spin-up period), extending from 1 January of 1990 to 31 December of 1999.

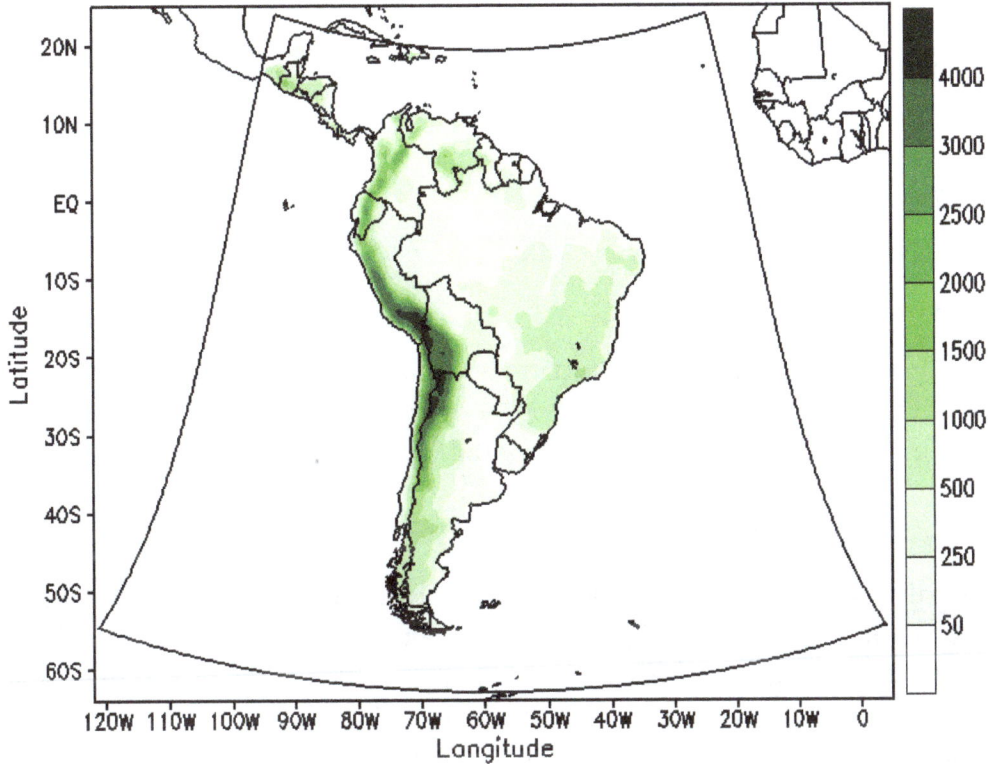

Figure 1. Model domain. Also shown is the topography of South America. Units, m.

2.1. Control experiment model

In the control experiment the model is forced using the ERA-Interim reanalysis data [26]. The greenhouse gas concentration corresponds to the present-day conditions. The distribution of aridity is obtained using the Budyko radiative dryness [21] and the UNEP aridity index [22]. The Budyko index has been used in many studies of land-surface effects, climate change and biogeography [27, 28, 29 and many others]. The UNEP index was adopted by UNEP to produce a dryness map [22].

The Budyko index, AI_B, is defined as $AI_B = R/(LP)$, where R is the mean annual net radiation; P, the mean annual precipitation and L is the latent heat of evaporation. Thresholds for different climate regimes are defined as:

$0 < AI_B \leq 1$=humid (surplus moisture regime; steppe to forest vegetation)

$1 < AI_B \leq 2$=semi-humid (moderately insufficient moisture; savanna)

$2 < AI_B \leq 3$=semi-arid (insufficient moisture; semi-desert)

$AI_B > 3$=arid (very insufficient moisture; desert)

The UNEP index, AI_U, is defined by $AI_U = P/PET$, where P is the annual precipitation and PET is the annual potential evapotranspiration. P is provided by the model while PET is calculated using the formula of [30]. Thresholds for different climate regimes are:

$AI_U \geq 1$ = humid regime

$0.65 \leq AI_U < 1$ = dry land

$0.50 \leq AI_U < 0.65$ = dry sub-humid regime

$0.20 \leq AI_U < 0.50$ = semi-arid regime

$0.05 \leq AI_U < 0.20$ = arid regime

$AI_U < 0.05$ = hyper-arid regime

Results of [31] showed that in general the climate variables, such as temperature, precipitation and evaporation, and the distribution of aridity over South America using both the Budyko and UNEP indices, for the present-day climate are well simulated by the model.

2.1.1. Climate change experiment on deforestation

The biomes distribution over South America according to the vegetation types given by BATS1e is given in Fig. 2a. In the deforestation experiment the entire tropical forest zone is converted into short grass (Fig. 2b). So, all the characteristic parameters of the tropical forest are replaced by those from short grass conditions according to BATS1e. Though extreme, it is important to evaluate a scenario of a hypothetical complete Amazon deforestation. The extreme scenario of total deforestation is useful to provide insight into underlying physical principles of the functioning of the climate system. Although it is unlikely that deforestation will affect the entire Amazonian forest, the extreme scenario of total deforestation is useful to identify the sensitivity of the climate system to changes in the land surface properties. In this experiment the effects of deforestation in Amazonia on the regional climate and aridity is studied.

2.1.2. Surrogate climate change experiment including deforestation

In this experiment the effects of global warming is taken into account together with the deforestation in Amazonia. For this purpose the methodology for generating a surrogate climate change scenario with a RCM proposed by [20] is used. It consists of a uniform 3 K temperature increase and an attendant increase of specific humidity. In this scenario, the ERA-Interim dataset of temperature is increased by 3K throughout the atmospheric column and the sea surface temperature OISST dataset [32] are warmed by 3 K. The atmospheric greenhouse gases concentration of the sensitivity experiment is set to two times its present-day values. A global mean equilibrium surface temperature increase of 3 K corresponds approximately to a CO_2 equivalent concentration of 710 ppm [33].

The methodology for generating a surrogate climate change scenario is dynamically consistent and easy to incorporate in a RCM. The procedure can be applied to the study of the regional response to a pseudo-global warming with an accompanying increase of the

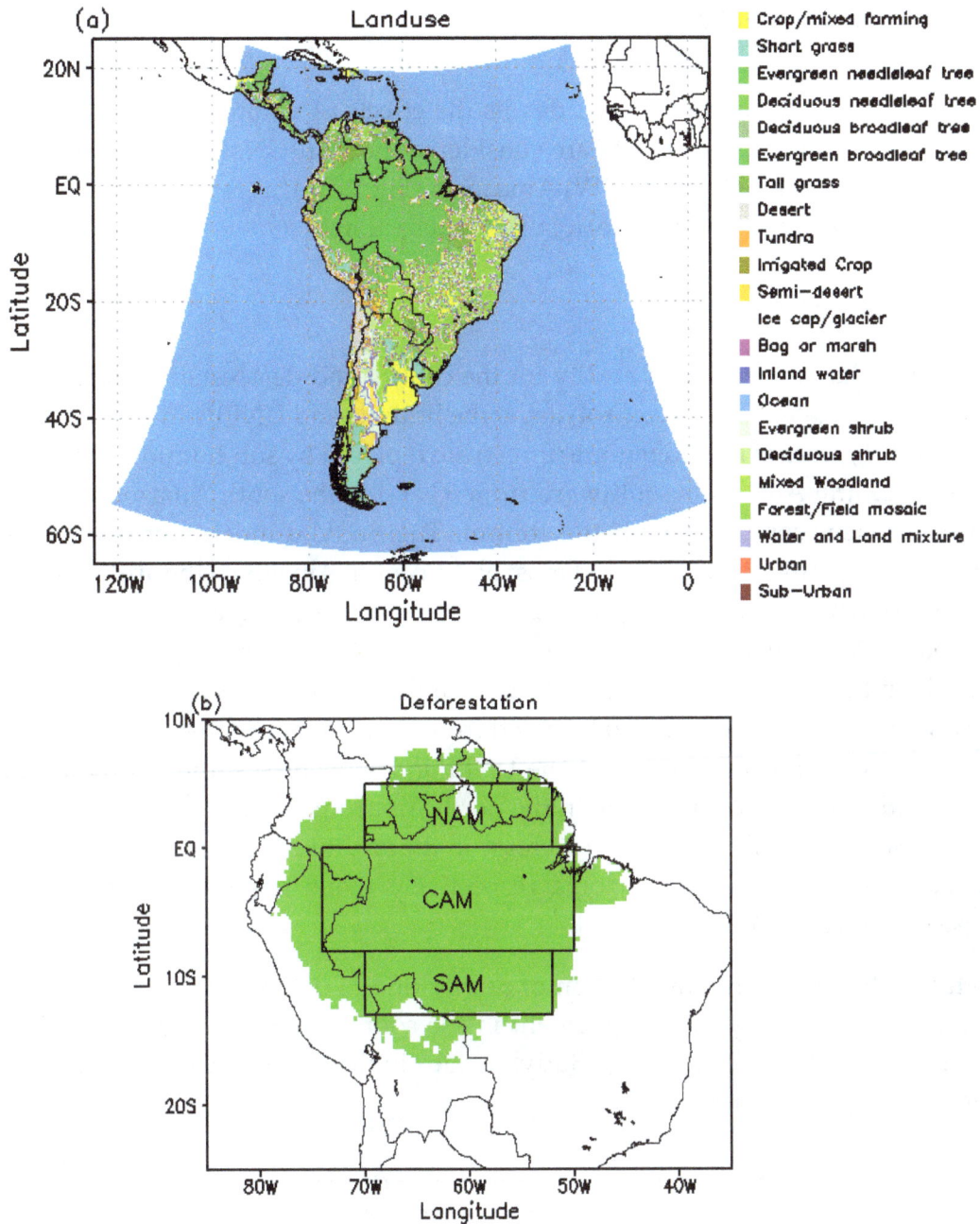

Figure 2. a) Vegetation types over South America according BATS1e; b) Region of Amazonia where the evergreen broadleaf trees are replaced by short grass in the deforestation experiment. Also shown are the areas denoting: north Amazonia (NAM), central Amazonia (CAM) and south Amazonia (SAM).

atmospheric water vapor content. However, the surrogate climate change scenario is only a sensitivity experiment and not a real climate change experiment. In a surrogate climate change scenario the response to a combination of a horizontally uniform thermodynamic modification of the initial and external fields plus an unmodified external flow evolution is studied. Otherwise a real climate change would be accompanied by changes in the planetary and synoptic-scale circulation. In spite of this drawback, the methodology allows us to examine certain processes in isolation [20, 34, 35].

3. Results and discussion

In order to discuss with more regional details the effects of deforestation and the pseudo-warming on Amazonia, three regions are considered: north (0-5N, 70W-52W), central (8S-0, 74W-50W) and south (13S-8S, 70W-52W) Amazonia (Fig. 2b). This is because the changes are different in these regions, as will be seen in the next sections.

3.1. Effect of deforestation

Figure 3 shows the distribution of aridity for the control and deforestation experiments and the change (deforestation minus control) using the Budyko and UNEP indices. As can be seen in Figs. 3a and 3b, areas of humid regime (forest) are replaced by sub-humid regime (savanna) in the part of central Amazonia southward from 5S and in the south Amazonia in the deforestation experiment compared with the control. The Budyko index increases (increase of aridity) in these regions. In the north and most of the central Amazonia the aridity is decreased (Fig. 3c). As shown in Table 1, taking into account the values of AI_B averaged over the entire three regions of Amazonia, the aridity increases 22% relative to the control in the south region. In the north and central areas there is a decrease of the aridity of 4% and 1.1%, respectively. For the case of the UNEP index, it can be noted from Figs. 3d and 3e that dry land substitutes regions of humid regime in Amazonia. The UNEP index decreases (the aridity increases) in the central and south Amazonia while in the north Amazonia it increases, as seen in Fig. 3f. These changes in the UNEP indicate an increase in the aridity of 22% and 4.8% relative to the control in the south and central Amazonia, respectively, while in the north Amazonia there is a decrease of 3% (Table 1).

Although the changes in the distribution of aridity due to deforestation using Budyko and UNEP indices show a very good agreement in the south and north Amazonia, the results diverge in the central region: the use of Budyko index indicates a decrease of aridity while the UNEP index suggests an increase.

Index Budyko	Region	I_B CTRL	I_B defor	AI_B defor minus CTRL	Change in I_B (defor relative to CTRL)	I_B (defor plus pseudo)	I_B (defor plus pseudo) minus CTRL	Change in I_B (defor plus pseudo) relative to CTRL
	North Amazonia	0.74	0.71	-0.03	-4%	0.89	+0.15	+20%
	Central Amazonia	0.92	0.93	-0.01	-1.1%	0.99	+0.07	+7.6%
	South Amazonia	1.00	1.22	+0.22	+22%	0.90	-0.10	-10%

UNEP	I_U CTRL	AI_U Defor	AI_U defor minus CTRL	Change in I_U (defor relative to CTRL)	AI_U (Defor + pseudo)	AI_U (Defor + pseudo) minus CTRL	Change in AI_U (defor + pseudo) relative to CTRL
North Amazonia	2.66	2.74	+0.08	+3%	1.66	-1.00	-37.6%
Central Amazonia	1.68	1.60	-0.08	-4.8%	1.21	-0.47	-28%
South Amazonia	1.36	1.06	-0.30	-22%	1.22	-0.14	-10.3%

Table 1. Values of AI_B and AI_U and the relative changes in the experiments of deforestation and deforestation plus pseudo-warming.

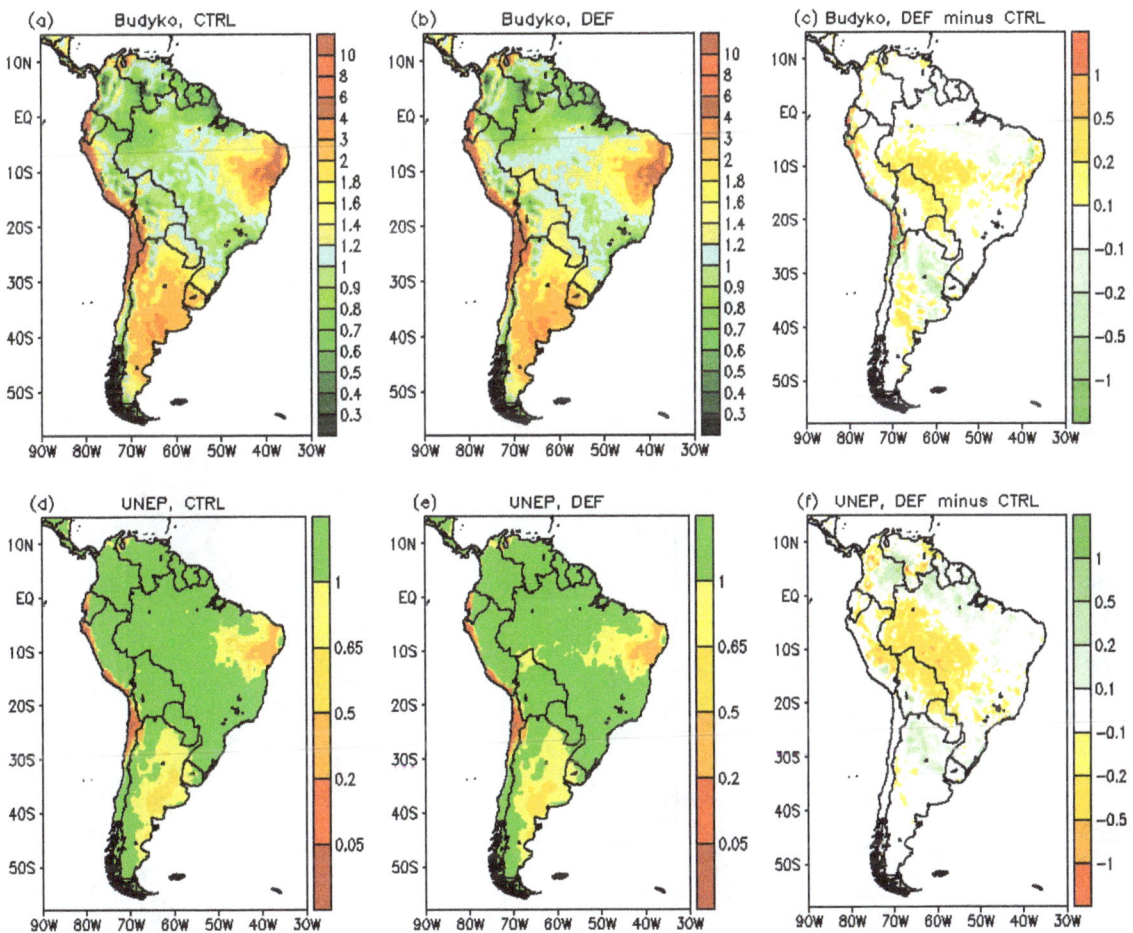

Figure 3. Distribution of aridity using Budyko index: a) control experiment, b) deforestation experiment and c) changes (deforestation minus control); and using UNEP index: d) control experiment, e) deforestation experiment and f) changes (deforestation minus control).

The changes (perturbed minus control) in the net surface radiation, precipitation, evapotranspiration and surface temperature due to deforestation are shown in Table 2. There is a decrease of the mean net surface radiation (-7.8 W m^{-2}) due to the increase of the land surface albedo; the mean evapotranspiration and precipitation decrease (-0.25 mm day^{-1} and -0.54 mm day^{-1}, respectively). The sign of the change in the surface temperature is different in the three regions of Amazonia. The mean surface temperature decreases in the north and central areas (-0.3C and -0.2C, respectively) and increases in the south region (+0.1C). As shown in Fig. 4a, the surface temperature increases by +0.6C in the south Amazonia and decreases by -0.9C in the north Amazonia. Since the higher decrease in evapotranspiration occurs in the south Amazonia it seems that the effect of the reduction in evapotranspiration in this region overcomes that of the increase of albedo while in the other two regions this does not occur. This leads to an increase of the temperature in the south Amazonia and a decrease in the north and central Amazonia. The changes in surface temperature in the three areas of Amazonia are in good agreement with the changes in the aridity given by Budyko index which indicates a high increase of the aridity in the south region (with a consequent increase in the surface temperature) while in the other two areas a decrease of aridity (and a consequent decrease in the surface temperature) is noted (Fig. 3c). The UNEP index also indicates a high increase of aridity in the south Amazonia and a decrease in the north Amazonia. However, differently from the Budyko index an increase of the aridity in the central region is noted.

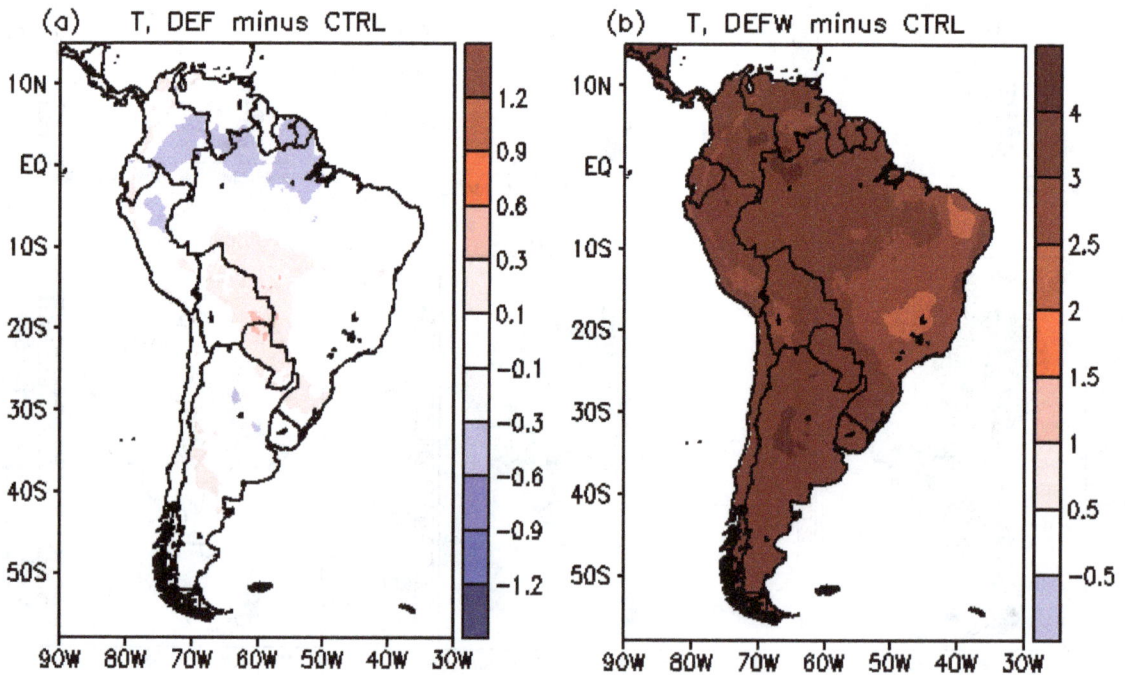

Figure 4. Changes in the surface temperature: a) deforestation minus control and b) deforestation plus pseudo-warming minus control. Units, ºC.

Experiment		Δ R (W m⁻²)	Δ P (mm day⁻¹)	Δ E (mm day⁻¹)	Δ T (C)
Deforestation	North Amazonia	-7.8	-0.10	-0.16	-0.3
	Central Amazonia	-7.8	-0.43	-0.24	-0.2
	South Amazonia	-7.7	-1.08	-0.36	+0.1
	Mean	-7.8	-0.54	-0.25	-0.1
Deforestation plus pseudo-global warming	North Amazonia	-1.7	-1.27	-0.48	+3.6
	Central Amazonia	-4.3	-0.57	-0.42	+3.6
	South Amazonia	-1.9	+0.53	-0.22	+3.3
	Mean	-2.6	-0.44	-0.37	+3.5

Table 2. Changes (perturbed minus control) in the surface net radiation (W m⁻²), precipitation (mm day⁻¹), evapotranspiration (mm day⁻¹) and surface temperature (°C) for the experiment of deforestation and deforestation plus pseudo-warming.

3.2. Effect of deforestation including pseudo-warming

Figure 5 shows the distribution of aridity for the experiment considering deforestation together with pseudo-warming and the change (deforestation plus pseudo-warming minus control) using the Budyko and UNEP indices. From Figs. 5a and 3b it can be seen that when the pseudo-warming scenario is taken into account the areas humid regime (forest) are replaced by semi-humid regime (savanna) northwards compared with the case of deforestation only. This leads to an increase of the aridity in this region. In the south Amazonia there is a decrease of the aridity, as shown in Fig. 5b. As can be seen in Table 1 the aridity increases 20% and 7.6% relative to the control in the north and central Amazonia, respectively, while in the case of only deforestation there is a decrease of aridity (4% and 1.1%, respectively). In the south Amazonia the aridity is decreased by 10% compared to the control while it increases in the case with only deforestation (22%).

Figures 5c and 5d show that in the case of the UNEP index there is a general increase of the aridity in the three regions in the deforestation plus pseudo-warming experiment compared with the control experiment. The increase of the aridity is higher in the north Amazonia (37.6%) followed by the central (28%) and south (10.3%) Amazonia. From Figs. 5d, 3f and Table 1 it can be seen that the aridity increases largely in the north Amazonia when the pseudo-warming is taken into account while it decreases in the case with only deforestation. Although in the two experiments there is an enhancement of the aridity in the central Amazonia the increase is much higher when the pseudo-warming is included. On the other hand the increase of the aridity in the south Amazonia is higher in the case of only deforestation.

It can be seen from above that the changes in the distribution of aridity due to deforestation together with pseudo-warming using Budyko and UNEP indices are in agreement. These

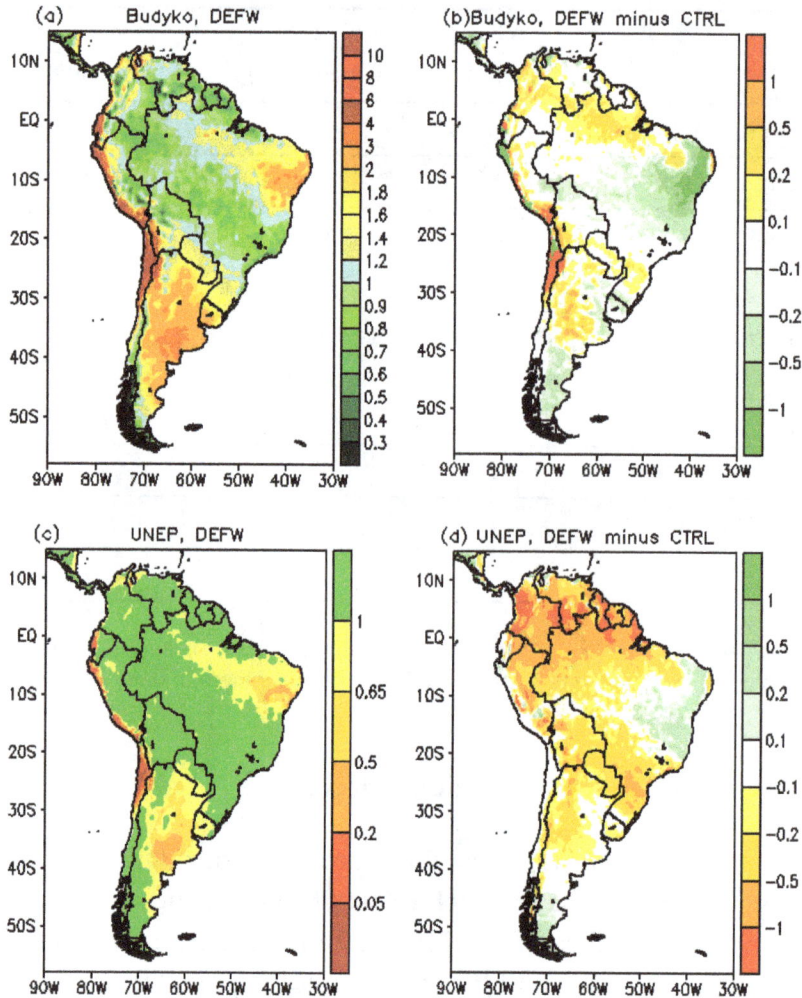

Figure 5. Distribution of aridity using Budyko index: a) deforestation plus pseudo-warming experiment and b) changes (deforestation plus pseudo-warming minus control); and using UNEP index: c) deforestation plus pseudo-warming experiment and d) changes (deforestation plus pseudo-warming minus control).

changes are higher compared to the case with only deforestation. On the other hand, the results diverge in the south Amazonia: the use of Budyko index indicates a decrease of aridity while the UNEP index suggests an increase.

Table 2 shows that the main changes in the Amazonia (an average over the three regions) are a warming of 3.5C and decreases in evapotranspiration (0.37 mm day^{-1}) and precipitation (0.44 mm day^{-1}) relative to the control. It can be seen from Table 2 that the inclusion of the pseudo-warming largely increases the changes in the surface temperature due to deforestation. However, deforestation may have a significant effect locally. As seen in Figs. 4a and 4b, the changes in the surface temperature due to deforestation may reach+0.6C in the south Amazonia, which correspond to 15% of the higher changes when the pseudo-warming is included (+4C). The increase in the surface temperature when the pseudo-warming is taken into account together is due mainly to the lower reduction in the net surface radiation in addition to the higher reduction in evapotranspiration. The changes in the surface temperature are large in the three regions of Amazonia. These changes are in good agreement with the changes in the aridity given by the UNEP index which indicate an increase of the aridity (and consequent increase of the surface temperature) in the three regions compared to the control (Table 1). The increase of the aridity is higher in the north Amazonia followed by the central and south Amazonia in agreement with the change in the surface temperature in these regions. The Budyko index also shows a higher increase of the aridity in the north Amazonia followed by the central Amazonia. However, in the south Amazonia an increase of the aridity is noted.

The present results agree with some studies with GCMs [8, 9, 10, 11, 36, 37] and with simple mechanistic models [5, 6, 38] which suggest that tropical South America is a region where significant portions of rainforest may be replaced by savanna (grassland) in future due to the global warming. The results also showed that the warming due to deforestation may have important effect locally; on the other hand when the effect of the global warming is included, the change of tropical forest areas of Amazonia by savanna may be enhanced compared with the present climate. This reinforces the hypothesis that due to global warming the process of savannization of tropical forest of Amazonia can be accelerated.

4. Conclusions

In this paper the relative roles of the land surface degradation in Amazonia and global warming on the regional climate and aridity were investigated using the RegCM4 model. Two experiments were performed: 1) deforestation and 2) deforestation together with global warming. The distribution of the aridity over South America, particularly over the tropical region, was obtained using the dryness index of Budyko and the UNEP aridity index. The results showed that the deforestation may have large influence locally (15% of the warming when the pseudo-warming was included together). The higher increase of the surface temperature occurred in the south Amazonia (+0.6C) whereas in the north and central Amazonia a decrease of temperature was noted (higher decrease of-0.9C). The changes in the distribution of aridity due to deforestation using Budyko and UNEP indices showed a very good agreement. It was suggested that there was an increase of 22% in the drying in the south Amazonia and a decrease of 3%-4% in the north Amazonia.

When the pseudo-warming was taken into account the changes in surface temperature were largely enhanced in relation to the deforestation case and the warming occurred in the entire Amazonia (higher increase of+4C). The changes in the distribution of aridity using Budyko and UNEP indices were similar. The aridity increased in most of Amazonia compared to the deforestation case. The higher increase occurred in the north Amazonia (20% for the Budyko index and 37.6% for the UNEP index).

Thus, the present study indicated that the global warming may affect the distribution of aridity over the tropical region of Amazonia, where significant portions of rain forest may be replaced by nonforested areas and this corroborates the hypothesis that the process of savannization of the tropical forest of Amazonia can be accelerated in future.

Acknowledgements

Thanks are due to Dr. Erika Coppola and the ICTP group for providing the RegCM4 code. Thanks are also due to Dr. V. Brahmananda Rao for going through the manuscript.

Author details

Sergio H. Franchito*, J. P. R. Fernandez and David Pareja

*Address all correspondence to: sergio.franchito@cptec.inpe.br

Centro de Previsão de Tempo e Estudos Climáticos, CPTEC, Instituto Nacional de Pesquisas Espaciais, INPE, SP, Brazil

References

[1] Nobre C A, Sellers P J, Shukla J. Amazonian deforestation and regional climate change. Journal of Climate 1991; 4: 957– 988.

[2] Sampaio G, Nobre C A, Costa M H, Satyamurty P, Soares-Filho B S. Regional climate change over eastern Amazonia caused by pasture and soybean cropland expansion. Geophysical Research Letters 2007; 34: L17709.

[3] Medvigy D, Walko R L, Avissar R. Effects of deforestation on spatiotemporal precipitation in South America. Journal of Climate 2011; 24: 2147-2163.

[4] Varejão-Silva M A, Franchito S H, Rao V B. A coupled biosphere-atmosphere climate model suitable for use in climatic studies due to land surface alterations. Journal of Climate 1998; 11: 1749–1767.

[5] Franchito S H, Rao V B, Fernandez J P R. Tropical land savannization: impact of global warming. Theoretical and Applied Climatology 2012; 109: 73-79.

[6] Moraes E C, Franchito S H, Rao V B. Amazonian deforestation: impact of global warming on the energy balance and climate. Journal of Applied Meteorology and Climatology 2013; 52: 521-530.

[7] Gash, J H C, Nobre C A. Climatic effects of Amazonian deforestation: some results from ABRACOS. Bulletin of the American Meteorological Society 1977; 78: 823–830.

[8] Cox P M, Betts R A, Collins M, Harris P P, Huntingford C, Jones C D. Amazonian forest dieback under climate-carbon cycle projections for the 21st century. Theoretical and Applied Climatology 2004; 78: 137– 156, doi:10.1007/s00704-004-0049-4.

[9]   Betts R A, Cox P M, Collins M, Harris P P, Huntingford C, Jones C D. The role of ecosystem-atmosphere interactions in simulated Amazonian precipitation decrease and forest dieback under global climate warming. Theoretical and Applied Climatology 2004; 78: 157–175.

[10] Salazar R F, Nobre C A, Oyama M D. Climate change consequences on the biome distribution in tropical South America. Geophysical Research Letters 2007; 34: L09708, doi:10.1029/2007GL029695.

[11] Malhi Y, Aragão L O C, Galbraith D, Huntingford C, Fisher R, Zelazowski P, Stich S, McSweenney C, Meier P. Exploring likelihood and mechanism of a climate-change-induced dieback of the Amazon rainforest. Proc. Natl. Acad. Sci. USA. Special feature: Sustainability Science 2009. available at: http://www.pnas.org_doi_10.1073.pnas.0804619106.

[12] Maynard K, Royer J.-F. Effects of "realistic" land-cover change on a greenhouse-warmed African climate. Climate Dynamics 2004; 22: 343–358.

[13] Voldoire A. Quantifying the impact of future land-use changes against increases in GHG concentrations. Geophysical Research Letters 2006; 33: L04701, doi: 10.1029/2005GL024354.

[14] Feddema J J, Oleson K W, Bonan G B, Mearns L O, Buja L E, Meehl G A, Washington W M. The importance of land-cover change in simulating future climates. Science 2005; 310: 1674–1678.

[15] Giorgi F, Marinucci M R, Bates G T. Development of a second-generation regional climate model (RegCM2). Part I: Boundary-layer and radiative transfer process. Monthly Weather Review 1993; 121: 2794–2812.

[16] Giorgi F, Marinucci M R, Bates G T, Decanio G. Development of a second-generation regional climate model (RegCM2). Part II: Convective process and assimilation of lateral boundary conditions. Monthly Weather Review 1993; 121: 2814–2831.

[17] Roads J O, Chen S-C. Surface water and energy budgets in the NCEP regional spectral model. Journal of Geophysical Research 2000; 105: 29539–29549.

[18] Chen S-C, Wu M-C, Marshall S, Juang H-M, Roads J O. 2 x CO_2 eastern Asia regional responses in the RSM/CCM3 modelling system. Global and Planetary Change 2003; 37: 277–285.

[19] Giorgi F, Coppola E, Solmon F et al. RegCM4: model description and preliminary tests over multiple CORDEX domains. Climate Research 2012; 52, 7-29.

[20] Schar C, Christoph F, Lutthi D, Davies H C. Surrogate climate-change scenarios for regional climate models. Geophysical Research Letters 1996; 23: 669-672.

[21] Budyko M I. The Heat Balance of the Earth's Surface. U.S. Department of Commerce, Washington D.C. 1958; 259 pp, translated by N.A. Stepanova.

[22] UNEP 1992: World Atlas of Desertification. Edward Arnold, London, UK.

[23] Grell G A, Dudhia J, Stauffer D R. A description of the fifth generation Penn State/ NCAR Mesoscale Model (MM5). National Center for Atmospheric Research Technical Note NCAR/TN-398+STR, 1994, NCAR, Boulder, CO.

[24] da Rocha R P, Cuadra S V, Reboita M S, Kruger L F, Ambrizzi T, Krusche N. Effects of RegCM3 parameterizations on simulated rainy season over South America. Climate Research 2013; 52: 253-265.

[25] Giorgi F, Jones C, Asrar G. Addressing climate information needs at the regional level: the CORDEX framework. WMO Bulletin 2009; 58: 175-183.

[26] Dee D P and coauthors. The ERA-Interim reanalysis: configuration and performance of the data assimilation system. Quarterly Journal Royal Meteorological Society 2011; 137: 553-597.

[27] Arora V K. The use of the aridity index to assess climate change effect on annual runoff. Journal of Hydrology 2002; 265:164-177.

[28] Sun Y I, Yan X D, Xie D T. Analysing vegetation-climate interactions in China based on Budyko's indices. Research Science 2006; 28: 23-29 (in Chinese with English abstract).

[29] Gao X, Giorgi F. Increased aridity in the Mediterranean region under greenhouse gas forcing estimated from high resolution simulations with a regional climate model. Global and Planetary Change 2008; 62: 195-209.

[30] Thornthwaite C W. An approach toward a rational classification of climate. Geographic Review 1948; 38: 55–94.

[31] Pareja D. Sensitivity experiments over Brazil considering global warming scenarios using a regional climate model. MSc thesis. INPE-17254-TDI/2083. Sao Jose dos Campos 2013 (In Portuguese).

[32] Reynolds R W, Rayner N A, Smith T M, Stokes D C, Diane C, Wang W. An improved in situ and satellite SST analysis for climate. Journal of Climate 2002; 15: 1609-1625.

[33] Randall D A and coauthors. Climate models and their evaluation. In: Climate Change 2007: The physical Sciences Basis. Contribution of Working Group I to the Fourth Assessment Report of the Intergovernmental Panel on Climate Change. Cambridge University Press 2007.

[34] Im E, Coppola E, Giorgi F, Bi X. Local effects of climate change over the Alpine region: A study with a high resolution regional climate model with a surrogate climate change scenario. Geophysical Research Letters 2010; 37: L05704, doi: 10.1029/2009GL041801.

[35] Winter J M, Eltahir E A B. Modeling the hydroclimatology of the midwestern United States. Part 2: future climate. Climate Dynamics 2012; 38: 595-611.

[36] Scholze M, Knorr W, Arnell N W, Prentice I C. A climate change risk analysis for world ecosystems. Proc. Natl. Acad. Sci. U. S. A. 2006; 103(35), 13,116–13,120.

[37] Cook K H, Vizy K H. Effects of Twenty-First-Century Climate Change on the Amazon Rain Forest. Journal of. Climate 2008; 21: 542–560.

[38] Franchito S H, Rao V B, Moraes E C. Impact of global warming on the geobotanic zones: an experiment with a statistical-dynamical climate model. Climate Dynamics 2011; 37: 2021-2034.

7

Study of Impacts on Continuous Shrinkage of Arctic Sea & Sea Level Rise – Can Glaciers be Growing and Creating New Challenges to UK & USA?

Bharat Raj Singh and Onkar Singh

1. Introduction

Present state of environment and continuous occurrence of natural disasters has made it inevitable for the environmentalists and scientists to extensively study and carry out detailed analysis of the following threats faced by civilization across the entire globe:

- Fast shrinkage of the polar ice and by 2040, there will be no polar ice seen during summer.

- Fast rise in the Sea Level,

- Danger for species like: polar bear etc.

- Ice sheets, where it meets the Atlantic sea, that this area may be affected by cold waves, heavy snow falls and intense storms.

- Permafrost may create further warming which cannot be reversed.

It is evident that the entire Arctic tundra region is melting, the frozen layer of soil known as permafrost is the growing concern and is considered as a threat by the scientists. The permafrost that is formed due to the fossils of the plants is undergoing transformation of being thawed and decomposes under climatic change for the past tens of thousands of years.

The continuation of this process is sufficient enough for the releasing of the methane gas causing irreversible global warming. Northern Alaska, USA and some other Arctic regions show the phenomenon of Termokarsts, where the melted permafrost layer lead to the collapse of the ground to hollow.

Another consequence is Tundra fires. Studies show that Tundra fires are also being a factor of region warming. The alarming rates of these fires as noted by scientists suggest that the Arctic could turn into a lethal source of methane in not less than a decade.

Whole community of the scientists involved, in the research and fieldwork is helping us to understand the growing threat of melting permafrost in the crucial Arctic region.

According to Dr. Hansen, our planet is on a dangerous course of passing irreversible tipping points with disastrous consequences. The melting of permafrost in turn releases toxic methane gases, resulting into more warming of the atmosphere.

Thus, it is essential to act promptly to avoid further catastrophic warming and stabilize the planet on which all lives depend, as permafrost's melt is a potential source of runaway global warming.

In this paper, authors are focusing mainly on the shrinkage of the polar ice and its serious effects on humanities, especially in January to March in USA and UK as well as on the entire global lively hoods.

2. Global warming fast facts

It is a fact that Global warming is human-caused and it will continue for centuries even if greenhouse-gas emissions are stabilized as per the experts of the International Panel of Climate Change (IPCC) report-2007as shown in Fig.1. The human activities not only linked with human activity to Earth's warming temperatures but its continued effect is causing rise in the seas' level, more intense storms, heavy snow fall and a host of many other environmental maladies[1-9].

Speaking at a press briefing in Paris, France on February 2 2007, the executive director- Achim Steiner, United Nations Environment Programme stated that the climatic change on our planet is undergoing drastic change due to the "Fossil fuel use, agriculture, and land-use.

Still there are many myths that climate change is regular process and there is as such no effect of global warming which is responsible for climate change, intense storms, heavy snow fall or cold waves. Majority of people still say that the ice age likely to advance in the near future.

Let us believe this theory for a moment then questions comes:

i.   Whether the earth's solar orbit is shrinking that makes the earth closer to the Sun and causing rise in the temperature? If so, the rotation of the earth around the Sun i.e. 364.256 days should get reduced.

ii.   As per the above thought process, if the ice age is likely to advance then why intense storms affecting islands near the coastal areas; why cold waves are advancing towards plane from hilly glaciers; why the Arctic sea is shrinking and causing heavy snow fall experienced in the USA & UK in January 2014?

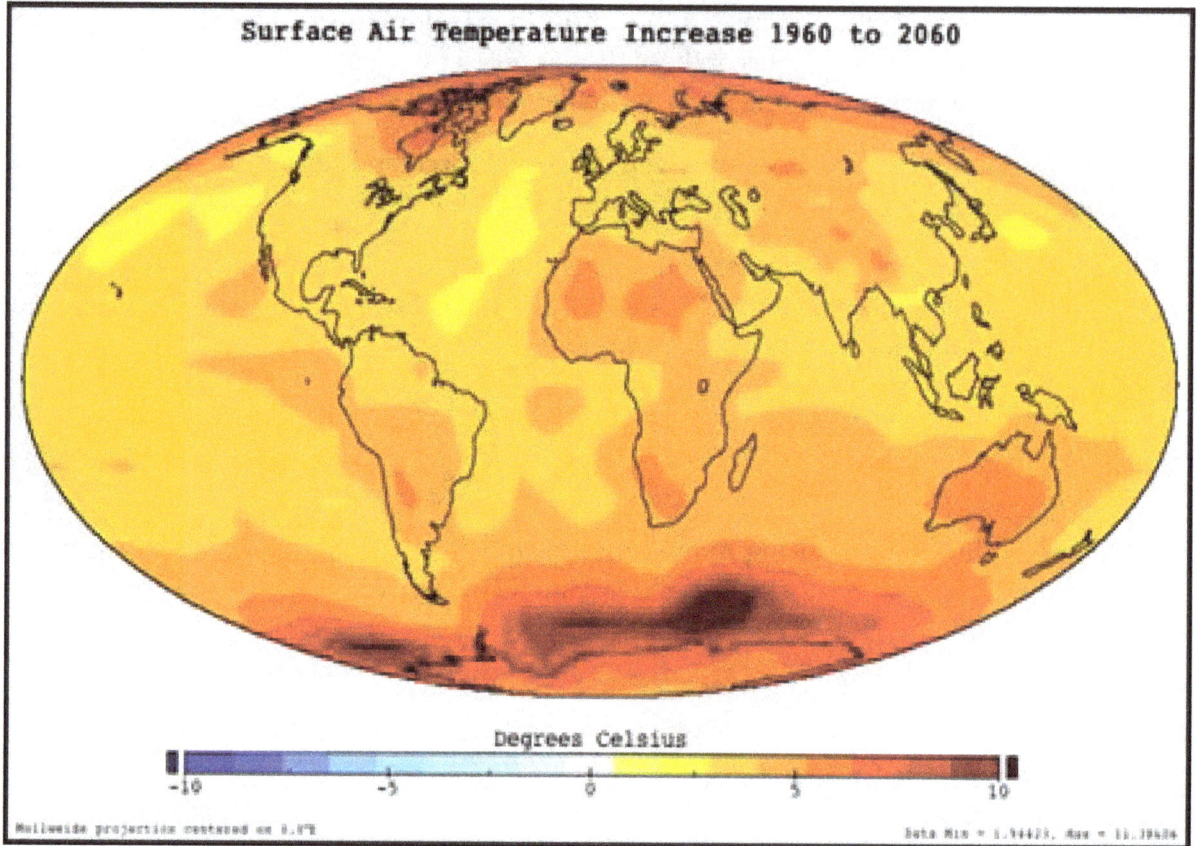

Source: Map courtesy NASA

Figure 1. A world map plots the annual average rise in the surface air temperatures from 1960 to 2060 based on the past measurements and future projections.

The truth remains that due to exploitation of the earth's resources and burning of huge quantity of fossil fuel, industrialization etc., global warming is now visible and damaging the climatic conditions, that is mainly caused by humans. This is further justified by global warming fast facts.

2.1. Is the change clear?

Actually-Yes, Our earth is showing clear changes in the climate: According to a study conducted by NASA's Goddard Institute for Space Studies- the average temperature has risen by 1.4 degrees Fahrenheit (0.8 degree Celsius) since 1880.

The last two decades of 20th Century had been the hottest in 400 years and it clearly shows the rate of warming. Moreover, the United Nations' Intergovernmental Panel on Climate Change (IPCC) reports that since 1850 among the last 12 years of climate; 11 years have been found the warmest.

In another very interesting study published by the Multinational Arctic Climate Impact Assessment Report of 2000 and 2004 showed that the average temperatures in Western Canada, Eastern Russia and Alaska have risen twice the global average temperature.

It may seem hypothetical but can be true that by 2040 the ice-caps at Arctic may not be seen because it is melting at rapid rate which can lead to a great threat to the habitation of the polar bears and other species of the region.

Studies show that there is rapid melting of the glaciers and mountain snows in the last few decades. To state is more clearly out of 150 odd glaciers in the Montana's Glacier National Park, only 27 glaciers are left to us since 1910. Furthermore in the Northern Hemisphere, it is shown that thaws are now a week earlier in spring and freezing about a week later.

According to the experts, due to this change, the Coral reefs that are very sensitive to small changes in water temperature suffered the worst bleaching because of the stress recorded in 1998. They have added to their conclusion that these events are to increase in frequency and intensity in the next 50 years with the rise in the sea temperature.

Wildfires, Heat waves and Strong tropical storms also attribute in part to climate change.

2.2. Factor behind it?

Yes, Burning fossil fuels and deforestation are its chief culprits. Burning of fossil fuels in transport as shown in Fig. 2, industrialization & power plants etc. and cutting of forests, known as deforestation are major contribution.

Figure 2. Burning fossil fuels through Transport Sector

2.2.1. Greenhouse gases trap heat

Owing to the excessive combustion of the fossil fuels, carbon dioxide (CO_2) and other greenhouse gases are released into the atmosphere, trapping heat and warming the atmosphere. CO_2 levels, today, are around 400 parts per million (*ppm*), that is 40% more than the highest natural level coming from coal and oil in the past 800,000 years, which varies from 180 to 300 ppm.

2.2.2. Losing forests makes it worse

Since deforestation is worldwide and trees are the major source for absorbing the excessive CO_2 from the atmosphere, therefore, concentration of CO_2 level in the atmosphere is on the hike that is allowing the heat to trap by making the atmosphere go warm.

Carbon dioxide, or CO_2, is the most dangerous greenhouse gas but, we cannot also ignore the contribution of Methane gas which damages the atmosphere equally to CO_2. As a matter of fact Methane's damaging percent to the atmosphere is less as it emits less CO_2 than coal. In fact, methane has the potential to undo much of the greenhouse gas benefits if shifted from coal to natural gas. An increase in the methane gas emissions is clear from quick expansion in natural gas development and an aging pipeline infrastructure. Methane emissions require us raising the bar on detection because whatever cannot be measured cannot be fixed. Low cost air pollution monitors are needed to be tapped for invention by the tech innovators to detect methane leaks in real-time [10-12]. Emissions of methane alone are not enough to study the cause of major changes in the climate: deforestation and the burning of fossil fuels such as coal.

2.2.3. So what do we do about it?

Keeping in view the rapid rate in the climate change; what we need to do is to come forward and learn about the dire consequences of the ecological imbalance and trying hands on recycling and buying local produce.

3. What sea-ice loss means for development in the arctic

Prediction for ice-free summers cannot be ruled out because of the shift of climatic cycle leading in the aggravation of global warming. Animals such as polar bears and walruses highly depend upon icy climate as shown in Fig.3.

Credit: World Wildlife Fund- August 23, 2012.

Figure 3. Threatening for polar bears that depend on the ice for habitat

On the contrary, ice shrinkage and melting has also paved way for increased shipping, tourism, oil, gas exploration and fishing too. It cannot be also denied that this change can pose challenge for the nations to grapple with the worst scenario to happen, as stated by a postdoctoral researcher during an address to the audience at Columbia University on Wednesday, September, 19, 2012.

Anne Siders researcher with the Columbia Center for Climate Change Law, one of the panel of researchers, discussed about the science as to how sea-ice de-freezing is taking place.

A predictably open Arctic Ocean creates opportunities and challenges for nations that ring the Arctic region. Here are some of them:

3.1. The opportunities

3.1.1. Fishing

According to the US National Oceanographic and Atmospheric Administration (NOAA), it can be said that rise in ocean temperature may create favorable conditions for the development of new commercial fisheries within the Arctic. Whereas, a U.S. plan of 2009 prohibits commercial fishing in U.S. waters.

3.1.2. More ship traffic

In terms of distance coverage through sea, thousands of miles of a trip could be cut when travelling along Northwest passage north of Canada or even the Northern sea route over Russia that a use to travel through the Panama Canal or the Suez Canal. Michael Byers, a professor of political science at the University of British Columbia made clear in an article in the Canadian newspaper "The Globe and Mail", that between 1906 and 2006, only 69 ships, in 2010, 18 and in 2011, 22 ships traveled through Northwest Passage. As per the Canadian Broadcasting Corporation reports, dwindling of the ice in the area has caused the trips to take place on cruise ships and private yachts.

3.1.3. Gas and oil

Offshore the Arctic region, according to the U.S. Geological Survey (USGS) in 2008, Arctic holds 90 billion barrels of oil, 1670 trillion cubic feet (47.3 trillion cubic meters) of natural gas and 44 billion barrels of natural gas liquids and a search for these resources is underway. On Monday, Sept. 17, 2012 the oil company Shell cleared that it has delayed the process of drilling the Alaskan Arctic coast until next year in reference to the stating situations.

3.2. The problems

3.2.1. Inadequate maps

Since the NOAA's maps and other navigational information is unavailable, it becomes difficult for the ships to ply through the route as the sea-ice is thick and impenetrable. Moreover, according to NOAA, the waters of Arctic that are charted were surveyed using obsolete technology dating back to the 1800s.

3.2.2. Little infrastructure or support

Siders say that though the coastline of Alaska is twice that of other 48 states, then too, there are limited resources regarding search, rescue operations and even to oil spill cleaning. One of the coast guard commandants of the region, Robert Papp, told to the media that, "we have almost no capability and in order to have a permanent availability of us, we need to do more investments on the infrastructure." It becomes necessary to bring in the example of a cruise ship, MV Clipper Adventure, which, in 2011, ran aground along the Northwest Passage and its passengers had to be rescued by the then coast guards.

3.2.3. Territorial disagreements

A CNN Money reports of the year 2008, said that far over the continental shelves, the Arctic coastal nations - the United States, Canada, Russia, Norway and Denmark (Greenland; a Danish territory) - are seeking to lay claim apart from the United States as it is not a party to the U.N. law of the sea treaty. Aligning the other territorial issues, Canada maintains that the Northwest Passage is sovereign Canadian territory, while U.S. and other nations advocates for international strait.

Siders adds, "there shall not be any more aggravating conditions, in terms of climate worsening, as that of the present in the Arctic region. The people are helpless but to work in spine chilling cold, dark and in the middle of storm."

4. Arctic summers ice free by 2040,

4.1. Study predicts

As per a recent study related to the global warming and its impact upon the sea-ice (see Fig. 4.) the summers in the Arctic Ocean may be ice free by 2040 when predicted through computer models while working on the greenhouse emissions and its impact especially by carbon dioxide (CO_2) from coal-fired plants and carbon mono-oxide by the automobiles.

The results of the computer study on the effect of the greenhouse gases on the atmosphere seemed alarming as it showed a steadily decline in the sea-ice level for decades and then abruptly disappearing.

Bruno Tremblay, an assistant professor of Atmospheric and Ocean Sciences at McGill University in Montreal, Canada that there are tipping points in the system and when we reach them things accelerate in a nonlinear way[14].

(Source: NASA satellite- by U.S. National Snow and Ice Data Center)

Figure 4. Satellite Images of Arctic Sea Ice in white on Sept. 16, 2012 and with the yellow line from 1979 to 2000, considering average.

4.2. How is It happening?

By September, the summer melt reduces ice cover to its minimum where with the arrival of the winters, it refreezes the sea-ice.

One of the simulation models study that in the last ten years, the ice coverage area has shrunk from about 2.3 million square mile (6 million square kilometers) to 770,000 square miles (2 million square kilometers).

Research & Studies according to the North American map shows that very small percent of sea ice shall be left along the north coastal regions of Green land and Canada. Whereas, the left of the ocean basin shall remain ice free throughout the summer.

Another, larger study shows that the winter ice shall defreeze from about 12 feet (3.7 metres) to three feet (one metre) thick.

Tremblay added to this study by saying that, "The oceans of Arctic do not absorb much of the sun's radiation for gets reflected into the space because the oceans at Arctic acts as giant mirror."

But it can also be stated that as the warmer average temperatures melt the ice, the mirror shrinks and so other part of the ocean creates further warming due to more absorption of the Sun's energy.

This ice shrink causes more warming due to more heat absorption.

"It goes into a positive feedback loop - a very efficient way of getting rid of the ice cover," as said by Tremblay.

Study of the climate models suggest that ocean circulation pattern can be altered due to global warming further driving warmer Atlantic waters into the Arctic.

"That is a positive feedback as well and it enhances the melting of the ice."

4.3. Serious consequences

Adding to the dire consequences regarding drastic changes in the region's climatic conditions, Tremblay suggests we could reach saturation point as we head towards a rise in temperature. That could happen at the latest by 2020 or 2030.

The dearth of constant cold conditions in the Arctic sea-ice might be perishable on animals such as polar bears that is the only habitation of it [15].

It would cause grave danger for the local residents to crave for their square meal, as they are mainly dependent upon fish as their staple food.

The only advantage left before us would be of transportation that would become feasible for the ships to sail [16].

5. Some recent winter storm studies

5.1. A big winter storm threatens U.S. and hit Canada with a cold snap

Thousands of commercial flights were canceled on Thursday, Jan 02, 2014, while a winter storm moves from the northeastern United States over a region with a 100 million [17]. While, Canada faces one of the worst cold snaps in recent winters with temperatures up to 41^0 degrees below zero in cities like Winnipeg in the Midwest, the 29^0 below zero in Toronto, the most populous city in Canada.

El portal FlightAware.com, which monitors air traffic, It is reported that by the local noon 2.233 flights within, to or from the United States suffered delays and 1.419 had been cancelled. The National Weather Service has warned that the winter storm "will cause serious disorders" remainder of week, with a mantle of snow and temperatures below freezing point across the region.

Is it expected to cause Blizzard Snow Storm tonight on Long Island, NY? New York authorities have indicated that this could block traffic on the Long Island Expressway, if driving conditions on the highway are too dangerous. It is also expected that in the city of New York snow accumulation reaches the 23 centimeters and in Albany to 35 inches, with temperatures between 26 and 35 degrees Celsius.

5.1.1. Heavy snowfall followed by very low temperature

The private weather service AccuWeather has indicated "very cold air was still affecting the upper Midwest, and "another wave of cold air hit parts of northeast after the blizzard with snow and could bring the coldest conditions in several years".

Very low temperatures followed by snowfall are "a brutal combination", commented meteorologist Tom Moore, of the Weather Channel TV station: "People who are in a vulnerable state will really suffer", Efe Ambrose reports. "It is a set of circumstances very, very dangerous", He added. Meanwhile Massachusetts Governor, Deval Patrick, who has authorized all state government employees to return home up to three p.m. on January 02, 2014.

Forecasters warned that a second wave of cold weather will affect the country on Sunday, particularly in the central area. On Sunday, January 05, 2014, Packers while playing in his field one of the matches of the final round of the National Football League at that time, En Green Bay in Wisconsin, temperature could fall from 28 degrees to below zero degree.

5.1.2. Cold wave in Canada

The storm is also affecting Canada, facing one of the worst cold snaps in recent winters with temperatures up 41 degrees below zero. From the central part of the country to the Atlantic coast, Thursday January 02, 2014, thermometer markings are at below 20 degrees below zero that, together with the effect called wind, put the wind chill below -30 degrees Celsius.

The authorities have warned that under these conditions, exposed parts of the body are likely to be frozen in minutes and advised for limiting spacewalks to a minimum.

In Northern Ontario, where there are large urban centers but if small indigenous communities, temperatures are reaching -50 degrees Celsius. En Montreal, the main city of the province of Québec, the temperature in the morning on January 02, 2014, was -38 degrees Celsius.

The utility of the province has requested that the energy consumption of appliances is reduced between 16.00 and 20.00 hours to cope with the expected extra demand due to low temperatures. In the province of Nova Scotia, on the Atlantic coast, forecasters predict that in addition to temperature extremes, the region will suffer a severe storm that deposited between 15 and 30 inches of snow in the next few hours.

En Toronto, where last week a 250,000 people went without electricity several days due to an ice storm, temperatures were at -29 Celsius with the wind effect. Although Toronto Hydro has ensured that all its customers have recovered the service after the ice storm, in some areas were taking intermittent outages due to the extra energy to fight the cold.Municipal authorities in Toronto have also enabled extra beds in shelters to accommodate more homeless. While, on the west coast of Canada, the situation is more moderate. The city of Vancouver is planning a maximum temperature today January 02, 2014, 8 degrees and likely to follow rains.

5.2. Midwest faced another round of snow, dangerous cold air

The Midwest is facing yet another snowstorm for the second half of the weekend with dangerously cold air to follow on *Sunday, January 5, 2014*, USA. The onslaught of the bone chilling cold breezes could develop in some areas.

The intensity of the snow shall be felt across the lower Great Lakes and in the Mid-Mississippi Valley through Sunday affecting Chicago, Detroit and St. Louis.

The rising of the storm shall also bring snow and slippery travel in Ohio and Tennessee valleys on Sunday morning to Sunday night (January5, 2014), with the rising of the ice concern in the Northwest.

From St. Louis to Chicago, Indianapolis, Detroit, Cleveland and London, Ontario the heaviest snow is forecast to fall.

Hit severely by the strong snow showers, dipping temperature and bone chilling freeze; transportation and travelling has become extremely difficult. The cities that had been caught by bone chilling storming winds, dipping temperature blended with heavy snow pour are: Memphis and Nashville, Ten; Louisville and Lexington, Ky.; Cincinnati and Columbia, Ohio; Charleston and Morgantown, W.Va.; Pittsburgh and Bradford, Pa.; Jamestown and Rochester, N.Y.; Toronto, Ottawa and Ontario I-40, I-64 and I-65 on Sunday night i.e., on January 5, 2014.

While addressing to the audience a sign of caution for severe conditions in Eastern Ohio to West Virginia and from Western Maryland to Western Pennsylvania, has been raised by a senior meteorologist Dale Mohler in Western New York on Monday & Tuesday [18].

Areas compressed with strong winds, plunging temperatures and heavy rainfall are likely to face whiteout conditions. It should be brought into the notice that the conditions in the climate are likely to be worsen in south of Buffalo, the New State and in the south of Watertown, Tug Hill region.

Mohler said that intense climatic conditions could close down major interstate highways, including I-79, I-80, I-81, I-90 and Route 219, for January 06-07, 2014. He added that there might be some possibility of some people being caught off guard and stranded by the storm due to dangerous cold wave blast and is also said that the new wave of frigid air shall reach the interstate highway I-95 Northeast on Monday, January 05, 2014.

Detroit experience afternoon highs just above zero Monday and Tuesday whereas, Chicago is not expected to exceed 10 degrees below on Monday.

During early February of 1996, it was the last time that Chicago faced such cold weather conditions where the temperatures remained below zero around the clock for a couple of days.

5.3. Storm and cold wave attack the northeast

A strengthening storm centered near Cincinnati, Ohio, on Mar 12, 2014, at 8 a.m., struck the New England coast tomorrow morning. Amazing temperature changes are occurring with this system. For example, at St Louis, the temperature was 80 ^{0}C at 4 p.m. yesterday.

This morning, 14 hours later, there was mix of snow and rain and it was 56. Heavy gusty thunderstorms announced the arrival of the cold front, and there could be a band of heavy showers and thunderstorms all the way to the Middle Atlantic coast tonight. Several inches of snow blanketed Chicago, and Buffalo will have blizzard conditions this afternoon and evening of Mar 12, 2014.

There will be a rapid freeze-up overnight from Boston to Washington, D.C., and temperatures will stay below freezing in all but the southernmost part of that area all day tomorrow. However, the high pressure area marking the center of the cold air mass will be east of Virginia by Friday, and the southwesterly flow behind it will take the cold air away quickly.

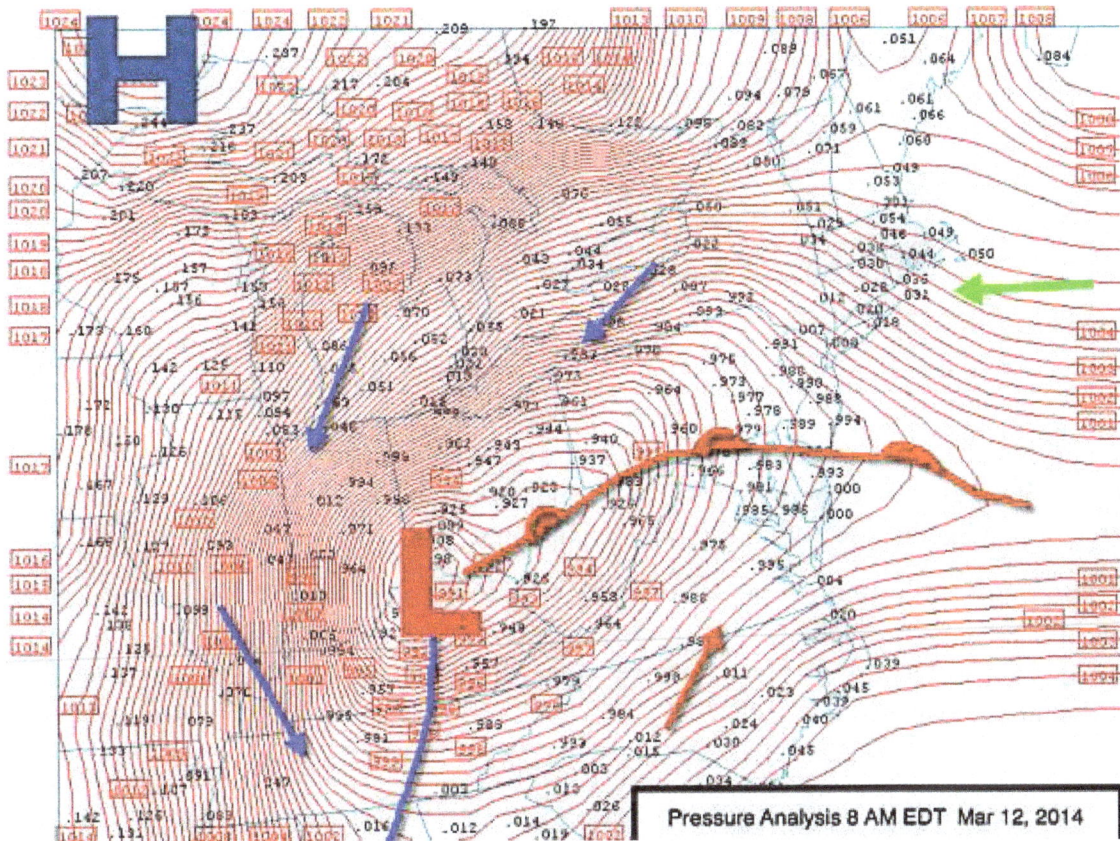

Source: AccuWeather.com

Figure 5. Pressure Analysis

This pressure map shows the storm center. The front to the east (re-line) marks the boundary between warm air to the south and progressively colder air to the north as shown in Fig.7.

5.4. Feel another cold wave passes more likely in January than March 2014

Across the Midwest and East the dipping temperature shall set the stage for heavy snow despite the official arrival of spring over the next several days feeling more or like January.

In continuation to this temperature conditions, this cold front has opened way for fresh arctic air to erase mild start to spring in the Midwest and East. Now the warmth of the weather conditions that of 50s in Boston, 60s in New York City and 70s in Washington-DC, on Saturday, March 29, 2014, seem disappear.

The typical January readings (in terms of average temperature, not what was recorded during this past frigid January 2014) elucidate the cold spell producing high and low temperatures at par.

Source: AccuWeather.com

Figure 6. Cold wave passes more likely in January than Monday, March 24, 2014 by Kristina Pydynowski

According to the weather source, the cold temperature during end of March 2014 shall be roughly not more than 15 to 20 degrees below normal in the Midwest and Northeast whereas in the Upper Midwest up to 25 degrees below normal.

By Wednesday, Duluth and International Falls, Minn., will experience at least two days of highs in the teens and subzero overnight lows this week. Highs in the 20s will return to Minneapolis, Detroit and Chicago. Temperatures will be held to the 30s southward to St. Louis and Cincinnati on Tuesday.

Figure 7. Lexington weather

March 24, 2014, Monday will be a cold day across the Northeast with highs in the teens and lower 20s across most of the St. Lawrence Valley, 20s southward to I-84 and 30s in Pittsburgh, New York City, Philadelphia and Atlantic City.

The latest arctic blast may set the stage for generally nuisance snow to spread across the Midwest Monday to Tuesday, while an Eastern New England is facing the threat of a blizzard at midweek.

In the wake of the blizzard, Wednesday will feel even colder than Monday across the Northeast as blustery winds howl on the storm's backside.

Outdoor spring sports and activities are ramping up at the collegiate and high school levels, and the cold threatens to cause problems for participants and spectators.

"The cold could force the cancellation or postponement of some scheduled events," stated AccuWeather.com Meteorologist Mike Doll.

Athletes will be able to put the winter jackets, extra layers of clothing, hats and gloves they will need to keep themselves warm over the next several days back into the closet later this week with another brief surge of warmth expected.

5.5. North American cold wave in the year 2013-14

The American Cold Wave (2013-2014) adversely affected Canada and the Eastern United States that extended from December 2013 to April 2014 with 2 episodic twists; first in December 2013 and the second in early 2014 caused by the southward shifts of the North Polar Vortex [19].

The first wave of record breaking cold air pushed into the Eastern U.S. that too before the temperatures receded to a moral stable range from December 6-10, 2013. A heavy snowfall (Fig. 7.) was recorded on January 2, 2014 by an Arctic cold front associated with the nor'easter that tracked across Canada and the United States resulting in the dipping of the temperature to unprecedented stages, and low temperature records were broken across the United States. Consequencing the closure of the roads, Business schools and cancellation of mass flights, with its effect [20-23]. In the same context, around 187 million residents of the continental United States were taken into its grip as the freezing cold air pushed in to the Eastern U.S. extending further towards the southern region from the Rocky Mountains to the Atlantic Ocean until it started receding from December 6-10, 2013 affecting more than 200 million people [24].

This happened due to the weakening of the Polar Vortex an abnormally cold trend swayed in the Eastern and Central United States, on December 1, 2013 from December 6 to 10, 2013.But, a sudden Stratospheric Warming (SSW) led to the breakdown of the regular Polar Vortex and subsequent southward movement of tropospheric Arctic air beginning on January 2. 2014, cutting down the cold rate [25].

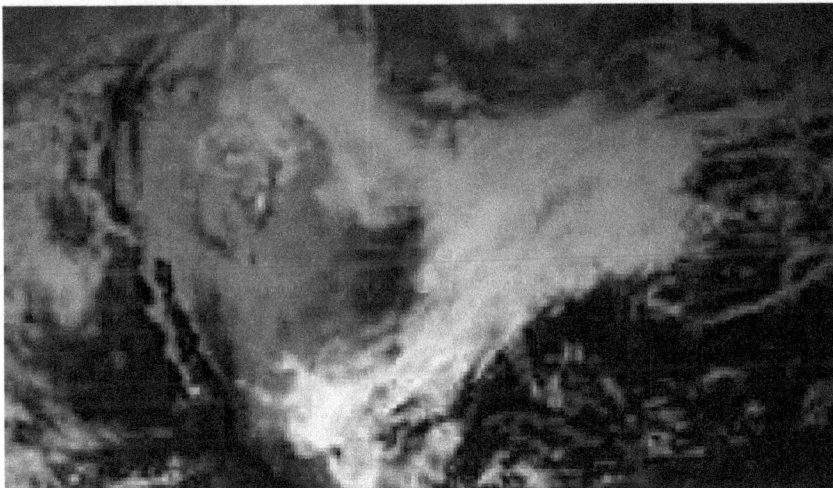

Source: Wikipedia

Figure 8. Satellite image of the severe winter weather took on January 2, 2014

As a result of the unusual contrast between cold air in Canada and mild winter temperatures in the United States, the jet stream deviated to the south (bringing cold air with it), as being stated by the UK Met Office. This led to the bitter wind chills and worsening the impacts of the record cold.

5.5.1. Recorded temperatures

- **United States of America**

The temperature at this stage on January 5, 2014, Green Bay, Wisconsin was -18 °F (-28 °C) whereas, the previous low recorded temperature was in 1979 [26].

Babbitt, Minnesota was recorded the coldest place in the country at -37 °F (-38 °C) on January 6, 2014. Owing to this, the bitter cold air reached near Dallas experiencing a low temperature of 16 °F (-9 °C).

On January 6, 2014, the low temperature recorded at O'Hare International Airport in Chicago was – 16 °F (-27 °C) where, the previous record was -14 °F (-26 °C) set in 1884 and tied in 1988 [27]. For the cold wave coverage in Chicago [28], the National Weather Service (NWS) adopted the Twitter hash tag # Chiberia (a portmanteau of Chicago and Siberia) [29-30]. Stiffing winds exceeded at 23 mph, when Chicago set its all time wind chill record of -82 °F (-63 °C) in 1983, but it could not be broken since NWS adopted a new wind chill formula in 2001 [31].

As compared to the average recorded temperature of last time i.e. below 18 in January 13, 1997; on January 6, 2014 it was recorded to be 17.9 °F (-7.8 °C) that is longest on record during a period of 17 years [32].

On January 6-7, 2014, Detroit hit a low temperature of -14 °F (-26 °C) and on January 7, 2014, at least 49 record lows for the day were set across the country [33]. On January 7, 2014, the high temperature of –1 °F (–18 °C) was only the sixth day in 140 years of records to have a subzero high [34]. On January 7, 2014, the temperature in Central Park in New York City was 4 °F (–16 °C). In 1896 for the day set, temperature was recorded low since its data collection initiated by the government [35]. On the other hand Pittsburgh bottomed out at –9 °F (–23 °C), setting a new record low on January 6–7, 2014 and Cleveland also set a record low on those dates at –11 °F (–24 °C). Temperatures in Atlanta fell to 6 °F (–14 °C), breaking the old record for January 7 of 10 °F (–12 °C) which was set in 1970. At Georgia, the temperatures fell to –6 °F (–21 °C) at Brasstown Bald, Georgia [36]. Tampa experienced a low of 34 °F (1 °C) in January 2014 after the cold air moderate and it reached even to the subtropical Florida.

- **Canada**

The eastern prairie provinces, Ontario, Quibec and the Northwest Territories that were the coldest parts of Canada experienced the hit of the weather. However, only Southern Ontario set temperature records.

Winnipeg was the coldest city in Canada during the hit of the early cold waves [37]. On January 5, 2014, the daily high in Saskatoon was –28.4 °C (–19.1 °F) with a wind chill of –46 °C (–51 °F)[38].

In Hamilton, Ontario a cold temperature record of -24 degrees C (-11 degrees F) was set. Whereas in London, Ontario it was -25 degrees (-13 degrees F) [39].

5.5.2. Related extreme weather

From the American Plains and Canadian Prairie Provinces to the East Coast a heavy snowfall or rainfall occurred on the leading edge of the weather pattern and due to the cold wind factor, strong winds prevailed throughout the freeze making the temperature feel at least ten degrees Fahrenheit colder than it actually was. Some places along the Great Lakes were also under wind warnings, in addition to the rainfall, snowfall, ice and blizzard warnings [40].

- **United States of America**

Boston experienced a temperature of 2 degrees F (-17 degrees C), on January 3, 2014 along with the wind chill and over 7 inches (180 mm) of snow. Whereas, Boxford, Massachusetts recorded a record snow fall of 23.8 inches (600 mm). Fort Wayne, Indiana had a record low of −10 °F (−23 °C) temperature. In Michigan, over 11 inches (280 mm) of snow fell outside Detroit and temperature around the state reached almost 0 °F (−18 °C). An official closure of government offices and school were announced after New Jersey had experienced a heavy snow fall of over 10 inches (250 mm) [41].

O'Hare and Midway Airports observed cancellation of their 1200 flights due to the snowfall between 5-7 inches (13 cm-180 mm) on January 5, 2014. [42]. Moreover, the Freezing rain caused a Delta Air Lines flight to skid off a taxiway and into a snowbank at John F. Kennedy International Airport, but with no injuries [43].

Further a fall of snow between 0.5 and 2 inches (1.3 and 5.1 cm) in Tennessee region and was lighter farther south [44].

A record breaking freezing cold with 4 degrees F (-16 degree C) on January 7, 2014, in 116 years, was recorded at New York city that came after days of unseasonably warm temperature, with daytime high dropping as much as 50 degrees Fahrenheit overnight [45].

- **Canada**

On January 5 and 6, 2014, in most parts of Canada, the front brought rain showers and snow event, which became the second nor'easter in nor'easter in less than a week in Nova Scotia and Newfoundland [46]. The onset of the front marked the end of this weather event bringing the bitterly cold temperatures with it. Due to lake-effect snow Southwestern Ontario also suffered again second time with heavy snow fall whole day on January 6 and 7 and part of January 8 [47], but Northwest Territories and Nunavut did not experience such chilling cold, but had a record-breaking blizzard on January 8, when further freezing cold towards south was coming to an end.

On other hand entire regions of Ontario and Quebec were undergone blizzard warnings [48]. Montreal has only suffered −24 °C with wind-chill factor overnight from January 6 to 7, when the same locations in Quebec have gone low temperature up to −34.5 °C on the night of January

2 to 3 and up to −41 °C on the night of December 25 to 26, which were also seemingly unrelated to the cold wave, but nearing records.

Canada experienced steady winds blow under the deep freeze around 30 to 40 kilometres per hour (19 to 25 mph). The winds reached 70 km/h (43 mph) with a gust as high as 100 km/h (62 mph) along the north shores of Lake Erie bringing local wind chill levels as low as to -48 degrees C (-54 degrees F) [49]. Several Ontario locations experienced cryoseisms or frost quakes along Lake Ontario and the St. Lawrence Valley [50].

- **Mexico**

A Tehuano wind event was created by the cold air that blew at 41 km (76 km/h; 47 mph) [51], from the Bay of Campeche into the Gulf of Mexico reaching Saltillo in the Northeast bringing temperature as low as -6 degrees Celsius (21 degrees F) [52].

1 centimeter (0.39 in) of snow grains were accumulated at Monterrey and registered a temperature of -1 degree C (30 degree F) [53].

5.5.3. Impact on electric power and transport

This extreme cold condition set for the cancellation of thousands of flights and seriously affected other form of transport. Many power companies in the affected areas asked their customers to conserve electricity.

- **United States**

Close to 100 daily snowfall records were broken across the northeastern, southeastern and south central United States, during the first episode of the cold wave, which extended from December 6–10, over 150 daily precipitation records [54] and [55]. Numerous airline flights were cancelled and there were reports of power outages [56-59].

Evan Gold of weather intelligence firm "Planalytics" called the storm and the low temperatures, the worst weather conditions event for the economy since Hurricane Sandy just over a year. It was not only nearly 200 million people were affected but calculated impact on Gold at $5 billion. Airlines were cancelled a total 20,000 flights after the storm began on January 2, 2014 with accumulated loss of $50 to $100 million [60]. Jet Blue airlines were adversely affected as its 80 percent flights go through New York City or Boston and were under major hit. Tony Madden, Federal Reserve Bank stated that as many schools closed, parents forced to remain home from work and vice-versa. He said that even people who could work from home, might not have done as much, excepting the insurance industry and government costs for salting roads, overtime and repairs [61-70].

Temperatures fell from 10 °C (50 °F) in the Middle Tennessee between January 5 and 6, 2014 dropping to a high of -13 °C (9 °F) on Monday, January 6, 2014 in Nashville. The power supply was interrupted leaving around 1200 customers to sit in dark in Nashville and around 7500 in Blount County [71]. The Tennessee Emergency Management Agency declared a state of emergency. Around 24,000 residents suffered power supply in Indiana, Illinois, and Missouri [72].

The power problem was reported by the Weather Channel in several states, abandoning cars on the highways in North Carolina and freezing rain in Louisiana [73].

• **Canada**

In late January 5, 2014 in Newfoundland, over 190,000 customers were left without electricity when the power supply could be restored by the next day [74].

Severe influence was on the air transportation which got delayed in Montreal and Ottawa. Some flights even got cancelled at Toronto Pearson International Airport due to low temperature and power cut [75-76]. ExpressJet, a partner of United Airlines, cancelled its flights into and out of Winnipeg, stating that the combination of extreme low temperatures and ice crystals exceeded safe operating guidelines for their airplanes [77].

5.5.4. Ecological impact

According to an opinion on the loss of ash trees, the spokesperson of the US Department of Agriculture in North America said that the extreme freezing conditions delayed the progressive loss of ash trees through it in North America [78]." The repudiation is based on scientific studies of under bark temperature tolerances of emerald ash borer in Canada [79-81].

5.6. New York City May 3, 2014 facing the dangers of 21st century great power war

World War I triggered with the assassination of Sarajevo. It was a beginning of the three decades of great power competition and warfare that culminated in the development and use of the atomic bomb. This was an era of militarized completion between rising and declining powers, intense disputes over territory and resources, arms racing, complex military alliances, rising nationalism and religious tensions. The beginning of this century, like the last, also is defined by deepening economic interdependence and competition, revolutionary advances in communications, and the belief that great power war would end civilization, as we know it, is thus, unthinkable. Yet from the Persian Gulf to the East China Sea there are more than enough wild cards to spark incidents that could spiral towards war.

Other cause of the war to begin was the pressure of the capitalist firms that have gained sufficient power in much of the world to write their own rules within economic framework. Intractable global economic crisis, with the actions essential to break the impasse thwarted by the extreme accumulation of wealth and power by elites determined to keep things as they are.

6. Results and discussion

From the above study, following points noticed that:

• In a span of ten years, sea ice coverage shrinked from about 2.3 million square miles (6 million square kilometers) to 770,000 square miles (2 million square kilometers).

- Expectation is that by 2040 or earlier the Arctic region may face ice-free summer since the Arctic ice is rapidly disappearing. Polar bears and indigenous creatures are already suffering from the sea-ice loss.

- Only the north coasts of Greenland and Canada shall remain covered with sea ice whereas rest of the ocean basin shall remain ice free through the summer (According to North America map).

- Glaciers and mountain snows are rapidly melting-for example, Montana's Glacier National Park now has only 27 glaciers, versus 150 in 1910.

- Winter ice will thaw from about 12 feet (3.7 meters) to 3 feet (1 meter) thick.

- Cold air in Canada and winter temperatures in the United States lead to bitter wind chills and worsening the impacts of the record cold temperatures on January 02, 2014.

- On January 06, 2014 Babbit, Minnisota was the coldest place in the country at (-) 37^0 F and cold reached to Dallas, experienced low temperature of (-)16^0 F.

- On January 03, 2014, Boston had a wind chill and over 7 inches snow whereas Boxford, Massachusetts recorded 23.8 inches of snow and schools and government offices were closed.

- In Canada, the major city Winnipeg was found coldest with temperature of −37 °C (−35 °F), on January 6, 2014 while it was −36 ° C (−33 ° F) on January 07, 2014, but it did not go above −25 °C (−13 °F) on both days.

- During the cold wave, there was heavy load on the power supply that got tripped and left nearly 1200 customers in Nashville and around 7500 customers in Blount County, without power supply. Between January 5 and 6, 2014, temperatures fell from 10 °C (50 °F) to -13 °C (9 °F) in Middle Tennessee, and the Tennessee Emergency Management Agency declared a state of emergency.

- Nearly 24,000 residents suffered power failure in Indiana, Illinois, and Missouri, USA.

- Owing to the power failure in Newfoundland, Canada, on January 5, 2014 around, 190,000 customers were left without power and air transportation was delayed at airports of Montreal and Ottawa. Almost all the flights were cancelled at Toronto Pearson International Airport due to the de-icing concerns.

7. Conclusions

From the study, it is found that Arctic Sea (Polar Ice) is shrinking very fast due to climate warming, particularly in warm summers. It is only on account of manmade effect of global warming. Thus, the following conclusions are listed below to act upon to fight with the dire consequences of fast shrinkage of Arctic sea, glaciers ice melt:

- There are very strong signs already been seen in the rate of sea ice change for the last one decade.

- Happening by 2020 or 2030 is not unrealistic while most of the ocean basin will remain ice free through the summer from the North America map.

- By 2040 only a small amount of sea ice will remain along the north coasts of Greenland, USA and Canada.

- The permafrost, that is a component for increase in the tundra fires, has the potential to turn the entire Arctic region from a vast carbon sink into a potentially lethal source of methane in less than a decade by enhancing the shrinkage of Arctic Sea.

- There are possibilities to grow glacier near north coasts due to heavy ice sheets meeting in the Atlantic sea.

- Ice sheets meeting to Sea-water may not convert quickly into water and create pressure drop, snow fall, extreme temperature drop to minus (-) 60-70 degree centigrade.

- USA & UK northern region may get affected with cold waves, disasters, intense storms, heavy snow falls and living life may not become conducive.

- The cold waves, extreme temperature drop may force living population in North American and Europeans to find new places for their living.

- Asian region especially India surrounded by three sides from sea and fourth side from Himalayan hills, may also affected badly with cold waves, disastrous intense storms, heavy snow falls nearby Himalayan glacier region; may cause heavy loss to the livelihood.

It is expected that the situation may go bad to burst every year and will continue in till next decade. During winter, New York, Britain and Canada i.e., northern belt, may suffer with extreme weather conditions such as: intense storm, heavy snow fall and power disruption.

Since the permafrost melt also confirms as a potential source of runaway global warming due to heavy methane availability, thus it is need of the hour to act very fast to help in stopping Climate Change due to Global Warming by adopting means to Save Earth and Save Life for happy living.

Acknowledgements

Enormous and extensive support provided by the management of School of Management Sciences, Lucknow, with a profound guidance under Harcourt Butler Technological Institute, Kanpur, has been a gliding way in accomplishing this project. Author Prof. Bharat Raj Singh expressed his special tribute to his mother Jagpatti Devi Singh who encouraged him to devote entire life for rendering the services to the human beings, though she had now left us for heavenly abode, leaving behind, her blessings for us.

Author details

Bharat Raj Singh[1*] and Onkar Singh[2]

*Address all correspondence to: brsinghlko@yahoo.com

1 Director, School of Management Sciences, Lucknow, Uttar Pradesh, India

2 Vice Chancellor, Madan Mohan Malviya University of Technology, Gorakhpur, Uttar Pradesh, India

References

[1] Archer, David. 2006. *Global Warming: Understanding the Forecast.* Hoboken, NJ: Wiley-Blackwell.

[2] Environmental Defence Fund, "Global Warming Myths and Facts," Accessed: January 17, 2009.

[3] Global Warming, Union of Concerned Scientists: Citizens and Scientists for Environmental Solutions. 2008. Accessed: November 27, 2008.

[4] IPCC Fourth Assessment Report: Climate Change 2007. Intergovernmental Panel on Climate Change. Accessed: November 21, 2008.

[5] Montague, Fred. 2006, Environmental Notebook: Observations, Principles, Trends, and Ideas about Life on Earth, Wanship, UT: Mountain Bear Ink.

[6] National Geographic News. "Global Warming Fast Facts." Accessed: January 14, 2009.

[7] Walker, Gabrielle and Sir David King. 2008. *The Hot Topic: What We Can Do about Global Warming.* Orlando, FL: Harcourt, Inc.

[8] Weart, Spencer R. 2003. *The Discovery of Global Warming.* Cambridge, MA: Harvard University Press.

[9] H. MacMillan, *Winds of change, 1914-1939* (London: 1966), 575; quoted in Michael Sherry, *The Rise of American Air Power: The Creation of Armageddon* (New Haven: 1987), 74.

[10] Methane studies fact sheet (Website: http://www.edf.org/sites/default/files/methane_studies_fact_sheet.pdf)

[11] Methane Study at University of Texas, FAQ page (web: http://www.edf.org/climate/methane-studies/faq).

[12] Methane Research Studies, Roster of Scientists (web: http://www.edf.org/climate/methane-studies/partners).

[13] Stefan Lovgren, Arctic Ice Levels at Record Low, May Keep Melting, Study Warns", National Geographic News, October 3, 2005.

[14] Bruno Tremblay, Journal *Geophysical Research Letters* -presented at the fall meeting of the American Geophysical Union in San Francisco, California, December 12, 2006.

[15] Sean Markey, "Polar Bears Suffering as Arctic Summers Come Earlier, Study Finds", National Geographic News, September 21, 2006.

[16] John Roach, National Geographic News, December 12, 2006, Codie Awards.

[17] Susana Romero, 2014, A big winter storm threatens U.S. and hit Canada with a cold snap, Jan 02, 2014(web: http://blog.susanaromeroweb.com/?p=6623&lang=en).

[18] Dale Mohler, 2014, "Blizzard conditions may develop from eastern Ohio to West Virginia, western Maryland, Western Pennsylvania, Before Its News, January 05, 2014, USA.

[19] Gutro, Rob. "Polar Vortex Enters Northern U.S.". Retrieved January 8, 2014.

[20] "N America weather: Polar vortex brings record temperatures". *BBC News – US & Canada* (BBC News Online). January 6, 2014. Retrieved January 6, 2014.

[21] Calamur Krishnadev (January 5, 2014). "'Polar Vortex' Brings Bitter Cold, Heavy Snow To U.S.". *The Two Way*. National Public Radio. Retrieved January 6, 2014.

[22] Preston, Jennifer (January 6, 2014). "'Polar Vortex' Brings Coldest Temperatures in Decades". *The Lede* (The New York Times). Retrieved January 6, 2014.

[23] "Arctic Monday for 140 million as 'POLAR VORTEX' barrels across the US: 4,400 flights canceled, schools closed as far south as ATLANTA and the coldest temperatures recorded in 20 years". *Daily Mail* (London). January 6, 2014. Retrieved January 7, 2014.

[24] Associated, The (January 7, 2013). "5 Things To Know About The Record-Breaking Freeze". NPR. Retrieved January 8, 2014.

[25] Spotts, Pete (January 6, 2014). "How frigid 'polar vortex' could be result of global warming (+video)". *The Christian Science Monitor*. Retrieved January 9, 2014.

[26] DeMarche, Edmund (January 4, 2014). "'Polar vortex' set to bring dangerous, record-breaking cold to much of US". *FoxNews.com* (Fox News). Retrieved January 6, 2014.

[27] "Chicago Gripped By Record Cold; Schools Closed, Public Transit Delayed". WBBM-TV. January 5, 2014. Retrieved January 8, 2014.

[28] "'It's too darn cold': Historic freeze brings rare danger warning". CNN. January 6, 2014. Retrieved January 8, 2014.

[29] "Metra plans on normal schedule for evening rush; CPS classes to resume Wednesday". *Chicago Sun-Times*. January 6, 2014. Retrieved January 8, 2014.

[30] "ChiBeria,' Chicago's biggest chill in nearly 20 years, headed our way". WLS (AM). January 5, 2014. Retrieved January 8, 2014.

[31] January 18, 2011 (2011-01-18). "Ask Tom why: What was the lowest wind chill ever recorded in Chicago?". Articles.chicagotribune.com. Retrieved 2014-01-15.

[32] Borenstein, Seth (January 10, 2014). "Weather wimps?". *Salisbury Post*. Associated Press. p. 1A.

[33] Doyle Rice (7 January 2014), List of record low temperatures set Tuesday USA Today

[34] Coldest Arctic Outbreak in Midwest, South Since the 1990s Wrapping Up, weather.com, Jon Erdman and Nick Wiltgen, January 8, 2014

[35] "Extreme Cold Wave Invades Eastern Half of U.S.". Weather Underground. January 7, 2014. Retrieved January 7, 2014.

[36] Record-setting cold turns deadly *Macon Telegraph* 7 January 2014

[37] Daily Data Report for January 2014 Canada government site

[38] "Daily Data". Climate.weather.gc.ca. November 12, 2013. Retrieved January 8, 2014.

[39] "At −24 C, Hamilton sets cold temperature record – Latest Hamilton news – CBC Hamilton". Cbc.ca. January 19, 1994. Retrieved January 8, 2014.

[40] "Alerts for: Simcoe – Delhi – Norfolk – Environment Canada". Weather.gc.ca. April 16, 2013. Retrieved January 7, 2014.

[41] "Snow, cold disrupt large swath of US; more to come". *Boston Globe*. Associated Press. 3 January 2014. Retrieved January 10, 2014.[dead link]

[42] "Winter storm 2014 strikes Midwest, Northeast; Snow storm 'Hercules' followed by 'polar vortex'". *WLS-TV*. 5 January 2014. Retrieved January 8, 2014.

[43] Fitzsimmons, Emma G. (January 5, 2014). "Jet Skids Into Snowbank at J.F.K. Airport". *New York Times*. Retrieved January 8, 2014.

[44] "Dangerously Cold Temperatures Settle Into Mid-State". WTVF News Channel 5. January 6, 2014. Retrieved January 6, 2014.

[45] Associated Press, The (January 7, 2013). "Record low temperatures in New York City". *Fox News*. Retrieved January 8, 2014.

[46] "Deep freeze extends from Winnipeg east to Newfoundland". Cbc.ca. 2014-01-02. Retrieved 2014-01-15.

[47] nurun.com (2014-01-07). "Schools closed, blizzard warning continued". Lfpress.com. Retrieved 2014-01-15.

[48] "Weather Alerts – Environment Canada". Weather.gc.ca. December 5, 2013. Retrieved January 7, 2014.

[49] North America's big freeze seen from space BBC News, January 9, 2014

[50] "'Frost quakes' wake Toronto residents on cold night – Toronto – CBC News". Cbc.ca. January 3, 2014. Retrieved January 7, 2014.

[51] Satellite Blog, CIMSS (January 3, 2014). "Tehuano wind event in the wake of a strong eastern US winter storm". *Space Science and Engineering Center*. University of Wisconsin-Madison. Retrieved January 8, 2014.

[52] "Synop report summary". Saltillo: Ogimet. Retrieved January 11, 2014.

[53] "76393: Monterrey, N. L. (Mexico)". *Professional information about meteorological conditions in the world*. Ogimet. Retrieved 30 January 2014.

[54] "U.S. Daily Precipitation Records set on December 6, 2013 | Extremes | National Climatic Data Center (NCDC)". Ncdc.noaa.gov. Retrieved 2013-12-08.

[55] "U.S. Daily Snowfall Records set on December 6, 2013 | Extremes | National Climatic Data Center (NCDC)". Ncdc.noaa.gov. Retrieved 2013-12-08.

[56] "Dallas, TX Weather Forecast from Weather Underground". Wunderground.com. Retrieved 2013-12-06.

[57] Donna Leinwand Leger and Doyle Rice, USA TODAY (December 7, 2013). "Snow, sleet and ice drive into eastern USA". Usatoday.com. Retrieved 2013-12-08.

[58] Jalelah Ahmed. "Winter storm halts travel; more snow forecast for Northeast | Al Jazeera America". America.aljazeera.com. Retrieved 2013-12-10.

[59] 21 hours ago. "Icy weather sends Texas' December power use to new record – Yahoo News". News.yahoo.com. Retrieved 2013-12-10.

[60] "Deep freeze may have cost economy about $5 billion, analysis shows". Durangoherald.com. 2014-01-10. Retrieved 2014-01-15.

[61] Karnowski, Stevework (9 January 2014). "Deep freeze may have cost economy about $5 billion". News & Observer. Associated Press. Retrieved January 10, 2014.

[62] "Today in Energy". U.S. Energy Information Administration. Retrieved 1 March 2014.

[63] Castellano, Anthony (January 3, 2013). "At Least 13 Died in Winter Storm That Dumped More Than 2 Feet of Snow Over Northeast". *ABC News*.

[64] "North America arctic blast creeps east". *BBC News*. January 7, 2014. Retrieved January 7, 2014.

[65] "OIA Can't Deice Frozen Jets". AviationPros.com. January 7, 2010. Retrieved January 7, 2014.

[66]   Kyle Arnold (January 6, 2014). "Sky Writer: Frozen aircraft fuel stalling nationwide air travel". Tulsa World. Retrieved January 13, 2014. "American Airlines said Monday that it's so cold in Chicago that airline fuel is freezing and they can't refuel planes. "Fuel and glycol supplies are frozen – at ORD (Chicago's O'Hare) and other airports in the Midwest and Northeast," said American Airlines spokesman Matt Miller."

[67]   "Service Alert". Amtrak. Retrieved January 7, 2014.

[68]   "Passengers stuck on Amtrak train 8 hours; Amtrak cancels some Chicago train service". WLS-TV. January 6, 2014. Retrieved January 8, 2014.

[69]   "500 passengers spend night on stranded Amtrak trains". Chicago Tribune. WGN-TV. January 7, 2014. Retrieved January 8, 2014.

[70]   Bone-Chilling Temps Shut Down Detroit People Mover, CBS News, January 7, 2014

[71]   Hayley Harmon (January 4, 2014). "Cold weather knocks out power for 7500 Blount County residents". WATE News. Retrieved January 9, 2014.

[72]   'Historic and life-threatening' freeze brings rare danger warning, CNN, January 6, 2014

[73]   "Winter Storm Pax Update: Hundreds of Thousands Lose Power, Drivers Abandoning Cars on Charlotte and Raleigh Highways". The Weather Channel. 12 February 2014. Retrieved 12 February 2014.

[74]   CTVNews.ca Staff (January 6, 2014). "Power restored to majority of customers in Newfoundland". CTV News. Bell Media. Retrieved January 6, 2014.

[75]   "Pearson airport delays: What you need to know". CBC News. January 7, 2014. Retrieved January 7, 2014.

[76]   "Extreme cold causes more delays, cancellations at Pearson". CP24. January 7, 2014. Retrieved January 8, 2014.

[77]   "Winnipeg deep freeze as cold as uninhabited planet – Manitoba – CBC News". Cbc.ca. Retrieved January 7, 2014.

[78]   Ziezulewicz, Geoff (January 6, 2014). "Cold weather could limit ash borer threat". Chicago Tribune. Retrieved January 8, 2014.

[79]   Purvis, Micheal (February 28, 2014). "Deer in danger from cold weather, but emerald ash borer likely not significantly affected, say scientists". Sault Star. Retrieved February 28, 2014.

[80]   Spears, Tom (February 4, 2014). "False hope: Deep freeze poses no threat to tree-munching emerald ash borer". Ottawa Citizen. Retrieved February 4, 2014.

[81]   Vermunt, Bradley; Cuddington, Kim; Sobek-Swant, Stephanie; Crosthwaite, Jill. "Cold temperature and emerald ash borer: Modelling the minimum under-bark temperature of ash trees in Canada". Ecological Modeling (Springer). 235–236: 19–25. doi: 10.1016/j.ecolmodel.2012.03.033. Retrieved 24 June 2012.

8

Dire Consequences on Little Shifting of the Earth's Spinning Angle – An Investigation Whether Polar Ice Shrinkage may be the Cause?

Bharat Raj Singh and Onkar Singh

1. Introduction

Environmentalist and Scientists are now of the opinion that the entire globe may face threats of: fast shrinkage of polar ice due to its melting and may eventually diminish by 2040, fast rise in the sea level, danger for species like: polar bears, penguins etc., northern portion of Canada, USA and UK may be affected by cold waves, heavy snow falls and storms due to shifting and melting of largest ice sheets in the Atlantic sea.

Scientists warn that the warming in the region of Arctic is due to the increment of Permafrost which is also one cause of the Tundra fires. The warming this way cannot be hence reversed and thus the entire Arctic region may turn into a dangerous source of methane from a vast carbon sink in less than a decade.

In view of likely disastrous implications, all the scientists involved, in the research and fieldwork are helping us to understand the growing threat of melting permafrost in the crucial Arctic region. Our Earth planet is on a dangerous course of passing irreversible tipping points with disastrous consequences due to the melting of green land, polar ice and permafrost which in turn releases toxic methane gases, resulting more warming of the atmosphere.

The future of sea level rise cannot be overruled by the ice sheets as they present alarming challenges in predicting their future response. It is calculated by using numerical modeling and as a result alternative approaches have been explored. A generalized approached is required in this matter to estimate their contribution to the sea level in the future.

In view of better identification and prediction of the melting and rising of the sea level a continuous monitoring via satellite is needed, according to the findings published in Nature

Figure 1. The Earth Planet

Geoscience. According to a survey and readings, the ice sheet covering Antarctica and Greenland contain about 99.5 percent of the earth's glacier ice that has the potential to raise the sea level by 63m (about 200 ft.), if melted completely

This entire action may lead to shift of heavy movement of masses of the Arctic sheets to sea and may likely to have an effect on the spinning angle of the earth due to differential changes in masses apart from the above mentioned threats.

2. Earth planet

Among the four largest terrestrial planets comes our earth. It is also the third planet from the Sun and also the densest one. We sometimes refer it as the Blue Planet [1] the Blue Marble, *Terra* or Gaia as shown in Fig.1. The genesis of our earth is estimated around four and a half billion years ago. The life on it initially appeared, as per the readings of the science, in the first billion years [2-5] in the oceans and began to affect its atmosphere and surface, promoting the spreading of aerobic as well as anaerobic organisms and causing the formation of the ozone layer. This ozone layer as well as the earth's magnetic field has the potential to trap the harmful ultra violet rays from the Sun from reaching the earth. Hence, life became possible to flourish on the land as well as in the water [6]. With the capacity of good physical properties and beneficial geological history, life is expanding and growing in its atmosphere.

Lithosphere, one of the geological features of the earth, is divided into several segments and tectonic plates that are formed over a period of many millions of years. Over 70% of the earth's surface is covered with water [7] and the remaining comprise of continents and islands having many lakes and other sources of water that contribute to the hydrosphere. *The earth's poles are mostly covered with thick sheets of ice that is the polar ice packs.* The inner of the earth is the thick layer of solid mantle, solid iron core and liquid outer core that generates the magnetic field.

The tilt of 23.4 degrees in the axis of the earth from its perpendicular of its orbital plane produce seasonal variations on the surface with one tropical year (365.24 solar days) [8]. During one orbit around the Sun, the earth rotates about its own axis 366.26 times creating 365.26 solar days or one sidereal year.

Owing to its feasible life generating conditions, it is a home to millions of species including humans [9]. The mineral resources and the biosphere products contribute much resources that are used to support a global human population [10].

2.1. Shape of the earth planet

An oblate spheroid, that is what the shape of the earth is. Means, it is flattened along the axis and bulged around the equator [11] causing the diameter to be 43 km larger than that of its poles [12]. The farthest point from the earth's centre is the Chimborazo volcano in Ecuador [13]. *The average diameter of the reference spheroid is about 12742 km, which is approximately 40,000 km/π, as the meter was originally defined as 1/10,000,000 of the distance from the equator to the North Pole through Paris, France* [14].

Since the earth has a tolerance of about one part in about 584, or 0.17%, the local topography deviates from this idealized spheroid only on small scale [15]. The Mount Everest (8848 m above sea level) has been attributed with the largest local deviations in the rocky surface of the Earth and the Mariana Trench (10911 m below local sea level) which together formed the equatorial bulge [16-18].

Some important physical and atmospheric characteristics of the Earth are shown in Table 1. From this data, it is evident that the Earth's radii through Polar and Equatorial are different such as 6356.8 km and 6378.1 km respectively. The polar radius is less than equatorial by 21.3 km or 43 km in diameter. The Earth's mass is 5.97219 x 1024 kgs. Circumferences through the equatorial and the meridional are 40075.017 km and 40007.86 km whereas the total surface area is 510072000 sq.km, out of which the land coverage is by 148940000 km² (29.2%) and the water coverage is 361132000 km² (70.8 %).

Geo-physical characteristics	
Mean radius	6371.0 km
Equatorial radius	6378.1 km
Polar radius	6356.8 km
Flattening	0.0033528
Circumference	40075.017 km (**equatorial**)

	40007.86 km (**meridional**)		
Surface area	510072000 km²		
	148940000 km² (29.2%) **land**		
	361132000 km² (70.8 %) **water**		
Volume of the Earth	1.08321×10¹² km³		
Mass	5.97219×10²⁴ kg		
	3.0×10⁻⁶ Suns		
Mean density	5.515 g/cm³		
Surface gravity	9.780327 m/s2		
	0.99732 g (**Earth gravity**)		
Moment of inertia factor	0.3307		
Escape velocity	11.186 km/s		
Equatorial rotation velocity	1,674.4 km/h (465.1 m/s)		
Axial tilt	23°26′21.4119″ (23.43°)		
Albedo	0.367 (**geometric**)		
	0.306 (**Bond**)		
Surface temp.	min	mean	max
Kelvin	184 K	288 K	330 K
Celsius	−89.2 °C	15 °C	56.7 °C
Atmosphere			
Surface pressure	101.325 kPa (at MSL)		
Composition	78.08% nitrogen (N₂) (**dry air**)		
	20.95% oxygen (O₂)		
	0.93% argon		
	0.039% carbon dioxide		
	About 1% water vapor (varies with climate)		

Table 1. Physical and Atmospheric Characteristics of Earth

3. Geomorphology

The surface of the earth, that comprise the action of wind, water, fire, ice and living things, is a combination of the landscapes and geological processes with chemical reactions that form soils and alter material properties and tectonic upliftment. The rate of change of topography under the force of gravity comes under the geological processes. As a matter of fact, the upliftment of the mountain ranges, the growth of volcanoes, isostatic changes in the land surface and formation of deep sedimentary basins are the result of geological processes. Thus, the earth's surface and its topography are an intersection of climatic, hydrologic and biologic action with geologic processes.

Much of the local climate is modified by topography, for example orographic precipitation which in turn change modifies the topography by changing the hydrologic regime in which it evolves. The intersection of the surface of the earth and the subsurface actions is well illustrated by the broad scale topographies. The geological process is responsible for the upliftment in the mountain belts. The sediments produced after husking of the high uplifted regions are transported and deposited elsewhere off the coast of the landscape [19]. The process of upliftment and deposition and of subsidence and erosion directly affect each other on progressively smaller scales at the involvement of the landforms. The loads of ice sheets, water and sediments can bring topographical changes through flexural isostasy.

Geographical cycle also named as the cycle of erosion is a model of broad scale landscape evolution developed by William Morris Davis between 1884 and 1889 which has been an eludication of the uniformitarianism theory first proposed by James Hutton (1726-1797). Opposing the Davis' model of single upliftment followed by decay, Walther Penck, in 1920, devised a model of cycle of erosion, since he thought that the landform evolution was better elaborated as an alternation between ongoing processes of upliftment and denudation. Since Penck's work could not be translated into English; his ideas could not be recognized for many years

The authors of early 19th century had tried their hands to attribute to the formation of the landscapes under local climate and to the specific effects of glaciations and periglacial processes. Significance of the genesis of the landscapes and the process of the earth's surface across different landscapes under various conditions has been tried and presented in a very well and on a more generalized way by Penck and Davis respectively.

3.1. Processes of geomorphology

The low temperature thermochronology, optically stimulated luminescence dating and cosmogenic radionuclide dating (geochronology) have enabled us to measure the rates at which geomorphic processes occur on geological timescales [20, 21]. Many advanced measurement techniques such as GPS, remotely sensed digital terrain models and laser scanning techniques have permitted quantification and study of geomorphological processes [22]. Further, with the help of modeling technique and computer simulation we know the working process of it.

The geomorphic processes generally listed into:

i.   the formation of regolith by weathering and erosion,

ii.   the transportation of that material, and

iii.   its deposition.

3.2. Glacial processes

Landscape changes happen due to glaciers because the movement of ice down a valley causes abrasion and plucking of the underlying rock and this abrasion further produces fine sediment that is termed as glacial flour. After the abrasion the debris transported by the glacier is called 'moraine'. The erosion of the glaciers are responsible for the formation of the U-shaped valley as seen in Fig.2, as opposed to the V-shaped valleys of fluvial origin [23].

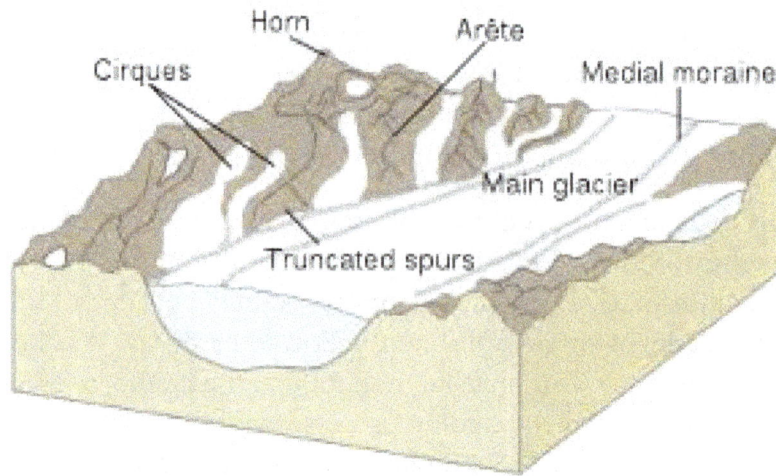

Figure 2. Features of a glacial landscape

Plio-Pleistocene landscape evolution is the one where glacial processes interact with other elements of landscape especially that of the hill slope and fluvial processes and is also responsible for sedimentation in high mountain environments. Elevated landscapes are the feature of recently glaciated environments as compared to those that have never been glaciated. In connection to this, 'paraglacial processes' that have been conditioned by past glaciation comes under 'Nonglacial geomorphic processes'. It contrasts with periglacial processes, which are directly driven by the formation or melting of ice or frost [24].

3.2.1. Glacial mass balance

The movement of a glacier works under the force of gravity. The flow of a glacier is downward because they collect mass (ice) due to the precipitation and wind pressure and get melted in small packs as shown in Fig. 3.

A glacier when in the state of equilibrium does not change in either in steepness or size because it works under the principle of accumulation = ablation. The equilibrium line altitude is because of zero net accumulation or ablation from the altitude. Advancement and recession in a glacier depends upon the changes in the rates of accumulation i.e. if the accumulation area of a glacier shrinks, for example, and the equilibrium line altitude rises, then the glacier will recede [25].

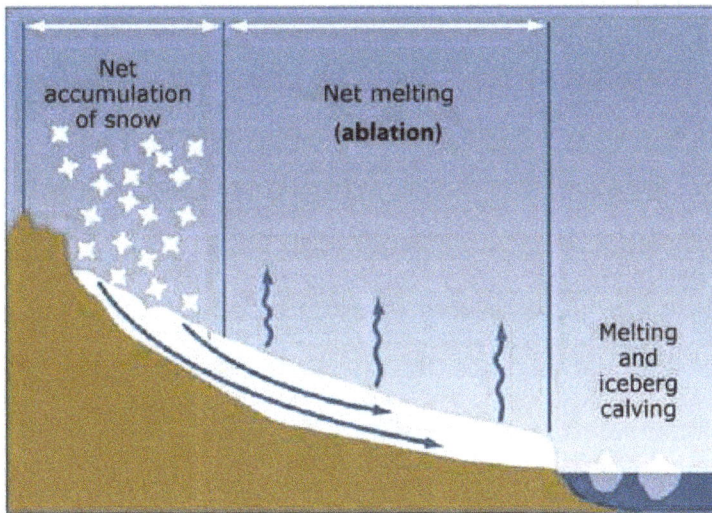

(Source: From the USGS Link: http://pubs.usgs.gov/fs/2009/3046/)

Figure 3. Components of mass balance of a glacier

The mass balance of a glacier is controlled by temperature and precipitation and is calculated by taking the difference between accumulation and ablation. If accumulation is greater than ablation, then the glacier has positive mass balance and will advance. If ablation is greater than accumulation, then the glacier has negative mass balance and will recede.

Gravity accompanied with the mass in a glacier pushes it to flow downward. A receding glacier flows slowly but flows in becoming thin with snout position receding backwards.

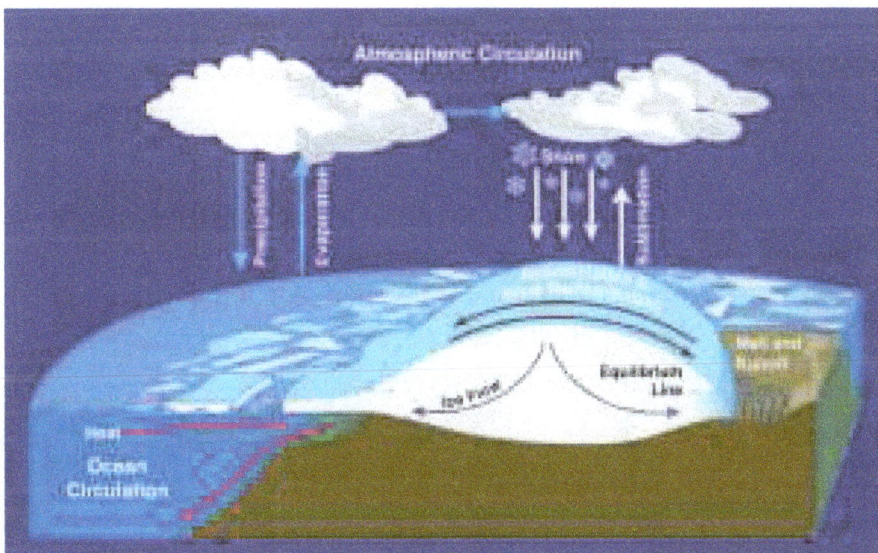

(Source: From Wikimedia Commons)

Figure 4. Glacier mass balance and atmospheric circulation by NASA.

The ice discharge from a glacier is by the accumulation and ablation area, thus maintaining a steady-state profile. The velocity of a flowing glacier is controlled by the glacier's mass. Some glaciers have dynamic flow driven by other factors, for example, surging glaciers, tidewater glaciers, ice streams or ice-shelf tributary glaciers.

3.2.2. Ice deformation and sliding

The process of deformation and sliding downwards is how the glaciers move (see Fig.4). The velocity, motion and flow of glacier is controlled under several factors as follow:

- Shape of geometrical formation of ice (thickness, steepness),

- Properties of ice in temperature and density,

- Geometrical valley,

- Conditions of bedrock such as: hard, soft, frozen or thawed bed,

- Hydrology in subglacial manner,

- Terrestrial environment like: land, sea, ice shelf, sea ice, and

- Mass balance in terms of rate of accumulation and ablation.

During the movement of the glaciers, there acts a driving stress. This driving stress, also called gravitational driving stress, is controlled by the density of ice, gravitational acceleration, temperature, ice thickness and ice surface slope. These resistive stresses basically operate at the glacier bed and make *basal drag* or *lateral drag* against the walls of the valley.

Three ways of the movement of glaciers under the driving stress:

i.   Creeping due to internal deformation

ii.   Sliding of basal

iii.   Subglacial deformation under soft bed.

All glaciers flow by creep, but only glaciers with water at their base (temperate or polythermal) have basal sliding, and only glaciers that lie on soft deformable beds have soft sediment deformation. If all three factors are present, one can have the ingredients to contribute to fast ice flow.

4. Ice melt fuels sea level rise concerns

4.1. Satellites monitoring of ice sheets for better prediction about sea level rise

Variations in the earth's gravitational field under changes in mass distribution with the movement of ice slabs into the oceans, is being detected by the satellites of Gravity Recovery and Climate Experiment (GRACE) since 2002. This provides us assistance and data as well, in monitoring the present condition of the ice sheets, at monthly intervals, under gravitational changes.

Dr. Bert Wouters, a visiting researcher at the university of Colorado, says that in the last few decades, as and when compared for the first few years of the GRACE mission; around 300 billion tonnes of ice is being lost by the ice sheets and its loosing rate is apparently increasing and adding substantial contribution in the sea level rise; almost double in recent years, compared to some few years.

The heavy loss of the ice sheets in the last few decades has not formed any general consensus among the scientists based on observations of the satellites. One agreement related to the loss of the ice sheets is that it may have been due to anthropogenic warming. In other words, natural processes such as severe fluctuations in atmospheric conditions, especially the shifting pressures in the North Atlantic, El Niño and La Niña effects, may have anthropogenic causes, as well as being due to the ocean currents.

"Dr Wouters state that, "if observations span only a few years, such 'ice sheet weather' may show up as an apparent speed-up of ice loss which would cancel out once more observations become available" [26].

The information received from the GRACE mission clearly mentions mass changes to the ice sheets after the comparison of nine years of data by a team of researchers. They detected that the ability to detect accurately an accelerating trend in mass loss depends on the length of the record.

Figure 5. The Satellite Monitoring about Sea Level Rise

The deformation in the ice sheets in the Antarctic region in the last few decades is alarming and its losses are unconvincing. If atmospheric fluctuations would be studied as the cause of the changing trend in the loss of the ice sheets at this region, it would leave a very meagre percent.

The satellite survey and study in the region specifically provides information about the mass loss of the ice sheets in the Antarctic region, and for Greenland, it will require us a time span of about ten years

It could be further added to our study, after the result of the satellite information regarding the mass loss in the ice sheets in the Antarctic region, that a continuous monitoring of the ice sheets through the satellite would be better to identify and predict the melting rate along with the observation in the sea level rise because 99.5 percent of the earth's glaciers of Antarctic and that of Greenland would raise global sea level to about 63m, if melted completely

The rational study regarding the sea level rise due to the ice sheets to 2100 might be 35cm if too high or low. Hence, prediction in the sea level rise, according to the studies, is an alarm for us to lift necessary steps to mitigate the onslaught.

4.2. Polar ice caps melt raises the oceans rise?

The rise rate of about half a degree celsius in the temperature in the last 100 years has no doubt caused Global Warming. Not to say, even half a degree would be enough to affect our planet' life. U.S. Environmental Agency (EPA) has stated in one of its survey report that, in the last 100 years, the sea level has raised from 6 to 8 inches (15 to 20cm) [27]. The rise in the temperature this way has paved the way for the melting of the polar ice sheets and floating icebergs to melt and could be if not on a large be one of the even small causes for the rise in the sea level.

(Source: by Tom Brakefield)

Figure 6. Antarctica accounts for about 90 percent of the world's ice

90 percent of the world's ice, around 2133 meters (7000 fts.) (including 70 percent of fresh water) is ice covered landmass in Antarctica at the south pole (Fig.6). It is hypothetical to say but, if the entire ice gets melt, the level in the seas around it would rise up to 61 meters (200 fts). Since, the average temperature in Antarctica is -37 degrees C, it cannot happen.

It can be said that the amount of ice covered at the Greenland, if gets de-freeze, would raise the level of sea around it to 7 meters (around 20 fts). On being close to the equator, the chances of de-freezing of the ice sheets is more as contrast to that of Antarctica.

The temperature variation of sea water has bigger impact over density of water. It is observed that water is most dense at 4 degrees celsius. The temperature above and below 4 °C, water density decreases and occupies a bigger space; leading to a proportionate rise in the water level in the oceans.

A report issued by the Intergovernmental Panel on Climate Changes of 1995, projected that by 2100, there would be rise of 50 centimeters (20 inches) with the lowest estimation of 15 centimeters (6 inches) and 95 centimeters (37 inches) the highest. This rise will be governed by the melting of glaciers and ice sheets along with the thermal expansion of the oceans. The rise of sea level at 20 inches cannot be considered trivial as far as the coastal regions are concerned, especially during storms, as it can bring havoc to the life and property nearly it.

4.3. New Greenland ice melt

The date of 'Nature Climate Change' of March 16th, 2014 (Sunday) revealed that there is rapid loss of ice sheets, over the past decade due to a rise in air and ocean temperature caused partly by climate change (see Fig.7). The increase in the melt has caused serious concern for the rise in the waters of the sea around the region even faster than projected, threatening the coastal life at large [28].

Shfaqat Khan, a senior researcher of Technical University of Denmark, wonders by saying, "North Greenland is very cold and dry, and is believed to be a very stable area. It is surprisingly to see ice loss in one of the coldest regions on the planet."

As of other glaciers on the island, the stability of the region is more important as it has much deeper attachments to the interior ice sheets. It is also said that, *"If the entire ice sheet were to melt -- which would take thousands of years in most climate change scenarios -- sea levels would rise up to 23 feet, catastrophically altering coastlines around the world."*

8 inches in the sea level rise has been observed globally since the start of 1900 and is projected to have further rise to 3 feet by the end of 2100.

Over three-quarters of the Greenland houses 680,000 cubic miles of ice sheet, stretching up to 3 miles in thickness in all directions finally confluencing at the sea nearby. The glaciers specifically of the southeast and the northwest have dumped enormous amounts of ice into the ocean, in the last 20 years and that has accounted for a more of 15 percent global rise of the sea level.

According to Dr. Shfaqat Khan, "These changes at the margin can affect the mass balance deep in the centre of the ice sheet". Moreover, the creeping rate of the sea levels is 3.2 *mm a year*, to

(Source: Henrik Egede-Lassen)

Figure 7. Helheim glacier in the southeast Greenland.

which Greenland contributes about *0.5 mm,* contrary to the real figure which is significantly higher. They calculate that between April 2003 and April 2012, the region was losing ice at a rate of 10 billion tonnes a year.

4.4. East Antarctic melting could raise sea levels by 10 to 13 feet

The study under Katie Valentine in 'Nature Climate Change' comprising 600 mile Wilkes Basin in the East Antarctica (Fig.8) states that, if the melts of ice would raise the sea level by 10 to 13 feet [29], it would be alarming and the researchers also find the region vulnerable because of the small 'ice plug' that may melt over the next few centuries. In this addition, East Antarctica could be a large contributor to the sea rise.

Figure 8. Study shows that East Antarctica region is perhaps more vulnerable in causing rise in the world's sea level for thousands of years. A study on Sunday, May 5, 2014 at 9:25 am.(Source: by Katie Valentine)

Matthias Mengel, a leading author in this study says that, "East Antarctica's Wilkes Basin is like a bottle on a slant. Once uncorked, it empties out. However, it is a distant threat." The authors conducting study on it says that warming can be limited to keep the plug in place. This was noted out when the method of simulation was adopted under scenarios with water found to be 1 to 2.5 degrees warmer than what it is today. We can however, on the basis of our observations, say that if, concrete steps are not taken, our planet would come under the hit of another 2 degrees Celsius ' hike adding much to the global warming. When giving a first look at the Wilkies Basin, we can conclude that the East Antarctic region might contribute to the sea level rise, although it is a talk of distant future [30-35].

Anders Levermann told to National Geographic that, "This is unstoppable when the plug is removed." Speed of its removal cannot be expected but, it's definitely a threshold.

No doubt, if the entire ice of Antarctica would melt, it would raise the sea level to about 188 feet. Another study of 2012 states that over the past decade, Antarctica has lost about 50 percent of the ice cover because Antarctic glaciers have started with irreversible melt which could lead the sea level to rise up to 1 centimeter. In one of the recent studies on the declining of the ice belt at Antarctica, it has come into view that the glaciers in this region have begun with a self-sustained retreat [36-46].

Adding to the study, Eric Steig said that, "These new results show that the degree of melting experienced by the Antarctic ice sheet can be highly dependent on climatic conditions occurring elsewhere on the planet."

In other very alarming study done by the National Snow and Ice Data Center in Boulder, Colorado, stated in March 2014 that, the region is experiencing the fifth-lowest winter sea-ice cover ever since 1978.

4.5. Passed point of no return of Antarctic glaciers

Two separate teams of scientists cleared that, the glaciers at Antarctica have passed a point of no return and will keep on melting rapidly especially that of the western Antarctica

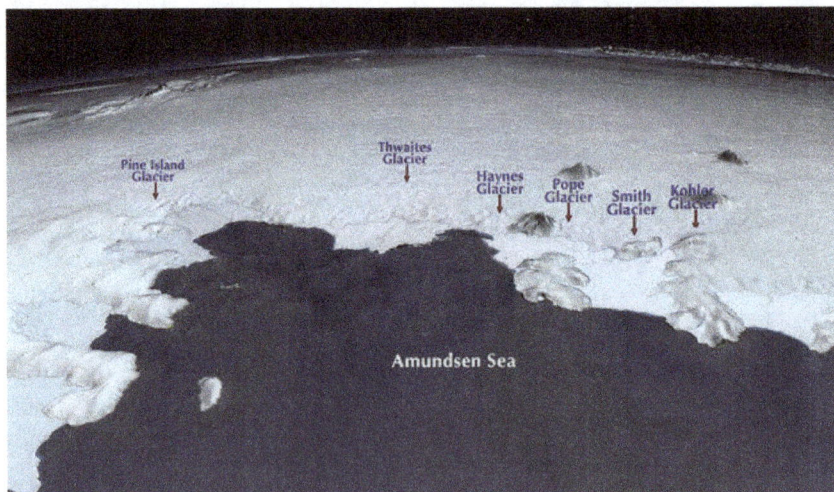

(Photo: NASA via AFP)

Figure 9. Getty Images by Traci Watson, on May 13, 2014; 9:14 a.m.

The likely result: a rise in global sea levels of 4 feet or more in the coming centuries, says research made public on May 12, 2014, Monday by scientists at the University of Washington, the University of California-Irvine and NASA's Jet Propulsion Laboratory as shown in Fig.9.

Sridhar Anandakrishnan, glaciologist of Pennsylvania State University says that, "It really is an amazingly distressing situation. This is a huge part of West Antarctica, and it seems to have been kicked over the edge."

Studies in progress show that the glaciers are in their stage of collapse and that is inevitable. We cannot reverse the situation. The Thwaites Glacier also known as 'the river of ice' is in its early stages of collapse and is almost inevitable. Half a dozen glaciers are dumping ice into the sea with pace which will give a rise of about 4 feet in the sea level, as per Eric Rignot, a glaciologist at the University of California-Irvine and laboratory at NASA's jet propulsion.

The same claim was made by Rignot at a briefing on May 12, 2014 Monday.

When studied about the retreating of the glaciers, Rignot and his team collected data made available to them through satellites and aircrafts to picture changes in six West Antarctic glaciers and the terrain underlying the massive ice, they found that the glaciers are stretching out and shrinking in volume by dumping mass of ice into the ocean.

Figure 10. Collapse of Thwaites Glacier

At the same time, the portion of each glacier projecting into the sea is being melted from below by warm ocean water as shown in Fig. 10. That leads to a vicious cycle of more thinning and faster flow, and the local terrain offers no barrier to the glaciers' retreat, the researchers report in an upcoming issue of Geophysical Research Letters.

A report in the mid of May 2014 Week's Science says the Thwaites Glacier will collapse, perhaps in 200 years. The paper doesn't specify the amount of sea-level rise associated with Thwaites' demise.

4.6. No way back for west Antarctic glaciers

The data of 19 years in the reported in the journal Geophysical Research Research Letters confirms that the melting of the West Antarctic glaciers are warming up in a speedy way in contrast to less warming of the southern hemisphere.

The West Antarctic ice sheet remained unstable and this had been an element of wonder among the glaciologists. According to the NASA research, there is enough water in the ice sheets of Amundsen Sea that is enough to raise the global sea levels by more than a meter. It is also said that, if the entire ice sheet of West Antarctic region changes to water, it can make the sea level to rise by at least five meters (See Fig.11).

4.6.1. Steady change

The recent study reveals that there is a steady change in the glacial grounding line, that clears us for the movement of the glacier towards the sea where its bottom leaves no abrasion on rock rather starts to float on water. Glacier has the nature to flow towards the sea and bear an iceberg that floats and later melts. Now, this has always been a matter of perplexity whether this process is going to accelerate?

The same is being considered by Eric Rignot, glaciologist at the Nasa Jet Propulsion Laboratory and the University of California, Irvine. Eric Rignot, glaciologist and his research partners estimates that the tidal movement can be responsible for bringing bending lines in the glaciers especially of the West Antarctic. This research was carried out when these glaciers were monitored between 1992 and 2011 on the basis of the data of European Space Agency. Since all the grounding lines had retreated from the sea by more than 30 kilometers. These are a bit hard to study because of their depth at which these are buried i.e. at more than hundreds of meters under the ice sheets.

An important clue can be obtained from the shift of the ice against the tidal waves and its flowing direction. It also signifies the acceleration of melting. It is also taken into consideration that the slow process of movement of the glacier cuts the rise in the sea level and as it inches towards the sea, more ice piles up behind it, collecting into mass.

4.6.2. Speeds up

As the water seeps under the ice sheet, it reduces friction rate and adds speed for the frozen water downstream and the whole glacier picks speed supporting the grounding line to move further upstream. Here, the melting could be slow at pace, but, not stoppable. The same phenomenon has been reported from the glaciers of Greenland

Photo: by NASA (*Source:* Earth Observatory via Wikimedia Commons).

Figure 11. Birth of an iceberg: a massive crack in West Antarctica's Pine Island glacier.

Prof. Rignot has again and again expressed his concern over the retreating of the glaciers of the West Antarctica in pointing out: "At current melt rates; these glaciers will be history within a few hundred years. We've passed the point of no return." So, the collapse of this sector appears inevitable.

5. Antarctic mass variation

We have derived multiple reasons for the mass de-freezing of the ice sheets under Antarctica and is thus giving pace for the ice under it to advance towards the ocean causing huge loss in the ice blocks.

5.1. Warmth in Antarctica

With the ice extension, Antarctica is also losing its ice cover. It seems amazing statement, but, analysis in this concern will certainly help us to find some concrete way in the context of global warming. The ice mass chart from GRACE satellite as shown in Fig.12 (a) and Fig.12 (b) helps us in this illustration.

On the basis of the observations and model studies, it is found that the sea ice is extending in the Antarctic sea despite the warmth in and around the Antarctica region. But, during each winter it allows to grow due to changes in ocean and wind circulation combined with changes in moisture levels. as compared to that of the Southern Hemisphere (SH), it remains cold and allow the ice to extent and grow. It can also be said that the growth of the Antarctica sea ice is likely because of the changes in the wind circulation combined with the moisture levels and that of the ocean currents. Moreover, the changes undergoing in the stratospheric ozone layer may also play significant role in this context.

Hypothetical studies observe its happening. Explanation by the studies of Zhang 2007 well contributes to understand the warming concepts in the Antarctic region [47].

An increase in the upper ocean temperature and a decrease in sea ice growth leading to decrease in salt rejection from ice in the upper ocean salinity and density are very clear by the model due to the increase in the surface air temperature and downward longwave radiation. The enhanced thermohaline stratification tends to suppress convective overturning along with the reduction in the salt rejection and upper ocean density, leads to a decrease in the upward ocean heat paving way for sea ice melting. The decrement in the ice melting from the ocean heat flux is faster as compared to the ice growth in the weakly stratified Southern Ocean, leading to an increase in the ice production. This mechanism is the main reason why the Antarctic sea ice has increased in spite of warming conditions both above and below during the period 1979–2004 and the extended period 1948–2004 [48, 49, 50].

The ice mass extends to grow in the Southern Hemisphere during winters more than the usual days. Whereas, the ice mass, at Antarctica, decreases during the summers, as per the satellite observations.

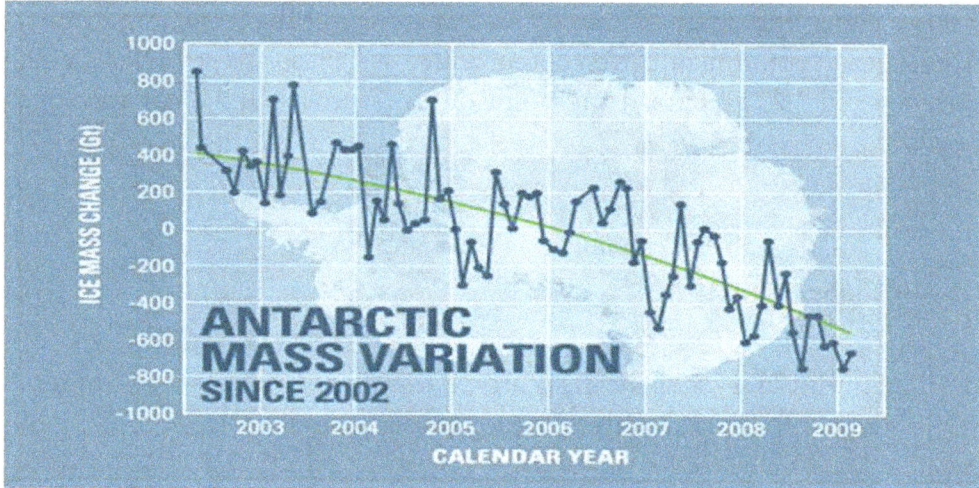

(a)

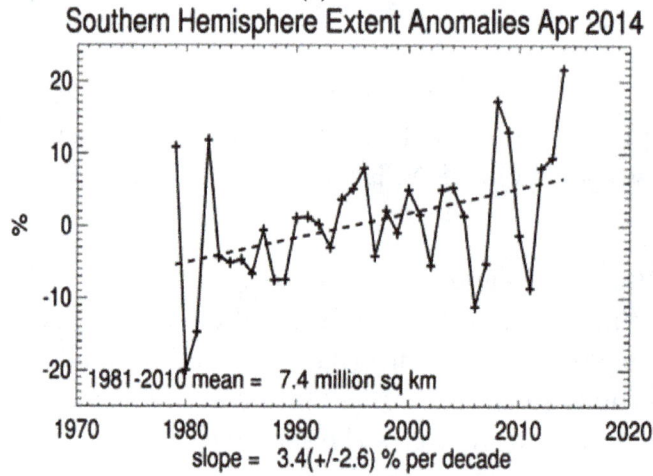

(b)

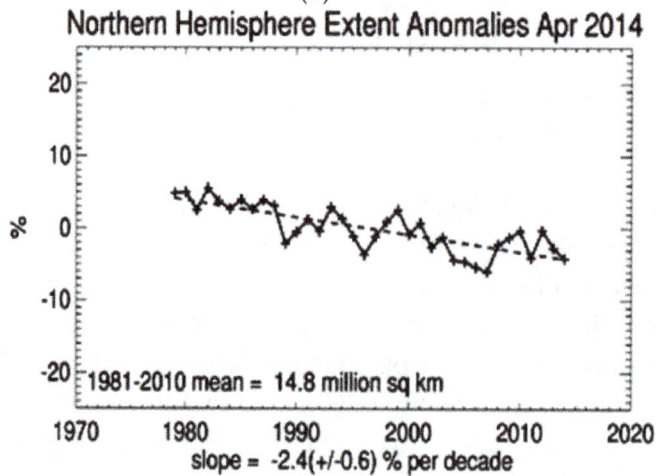

(c)

(Source: http://www.nasa.gov/topics/earth/features/20100108_Is_Antarctica_Melting.html; http://nsidc.org/data/seaice_index/; http://nsidc.org/data/seaice_index/)

Figure 12. (a). Antarctic Ice Mass Loss [manual update]; (b). Antarctic Ice Extent Increase updates annually; (c). Arctic Ice Extent updates annually

It can be concluded, after these kind of researches, that more snow precipitation might be expected in the Antarctica in the future than compared to today's scenario where the ice discharge rate has increased.

This is confirmed by the following data:

- Warming of Antarctica

- Increase in the sea ice of Antarctic

- Decrease in the ice mass of Antarctic land

5.2. Warm Arctic: Causes and concern

Since the Arctic (Northern Hemisphere) acts in the opposite direction regarding ice extent with losing ice mass as shown in Fig.12(c), it is not gaining that much of ice extent as that of the Antarctica. We may say that, because Northern Hemisphere has more land surface than that of the Southern Hemisphere which is mostly water body and ice mass, the two hemisphere acts in two opposite ways.

6. Results and discussion

From the above study, it is seen that the Earth's land [51-53] covers by 148940000 km^2 (29.2%) and water by 361132000 km^2 (70.8 %) that means every mm rise in sea will have a shift of melting ice into water around:

a.   Assuming average sea level rise of 0.5mm, then additional water will be added to sea water as under:

- Area of Water=361132000 km^{2}x 1000 x1000=361132x10^9 sqm per year

- Rise of water=0.5mm/1000=5x10^{-4} m

- Total water added=(361132x10^9 sqm) x (5x10^{-4} m) =1805660x10^5 cum (i.e.180.566 billion tonne)

b.   By year 2100, when sea rise is likely to be raised to 3.6 feet=1.1 Meter, then additional water will be added:

- Multiplier = (1.1m x 1000/0.5mm)=2200

- Water will be added=(180.566 billion tonne x 2200)= 397.245 trillion tonne

Thus, such heavy weight shift of (approx.) 400 trillion tonnes minimum or (approx.) 1200-1450 trillion tonnes maximum from polar ice-sea or Northern / Southern coast and green land to sea water, might force to the change in spinning angle of the earth from 23.43 degree to 23.43^0 (+ or -) as seen in Fig. 13. Further detailed analysis is still required or model is to be prepared to find out the exact date and time as to when such situation may arise.

What would be the fate of the Earth and its living creatures, when it happens?

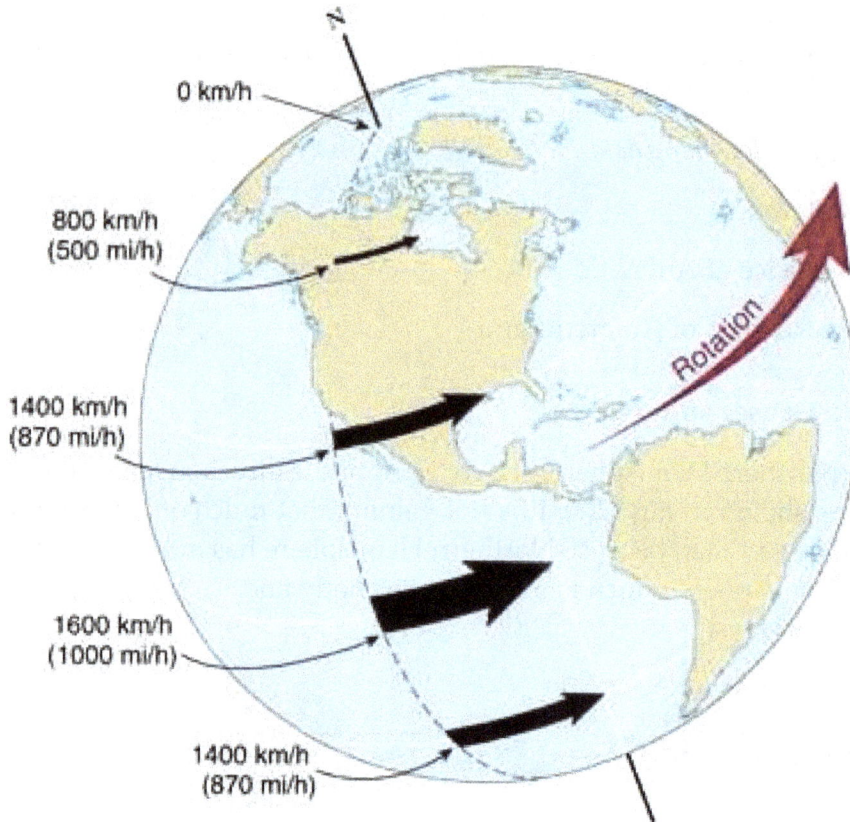

Figure 13. Earth Rotational Angle 23.43⁰

7. Conclusions

From the study, it is very much clear that Global Warming is happening and Polar Ice melt / Green land ice melt is continuing fast. This will not only affect our living and developments but have dire impacts on:

- The poles of the earth are patched completely with solid ice that of Antarctica and the sea ice.

- Fast shrinkage of the polar ice will diminish by 2040, fast rise in the Sea Level, danger for species like: polar bears, penguins etc., northern portion of Canada, USA and UK may be affected by cold waves, heavy snow falls and storms due to shifting and melting of heaviest ice sheets in the Atlantic sea. Permafrost may create further warming which cannot be reversed.

- Tectonic movement, wind action, ice, fire, and the living things on the surface with the interference of the atmospheric pressure and temperature combines to make geological structure of the earth and its processes.

- In the landscape change, glaciers play wide role. A gradual movement of ice down the valley causes scraping and sculpting of the underlying rocks producing sediments in the form of glacial flour.

- In the last 100 odd years the sea level has risen to about 6 and 8 inches (15 to 20 cm) due to global warming as being stated by the U.S. Environmental Protection Agency (EPA).

- To an average depth of 2,133 meters (7000 fts), the region of Antarctica is covered with ice and if the entire ice of the region would melt then, the rise in the sea level around it would come to 61 meters (200 fts). Since, the average temperature of Antarctica region is -37 °C, there is no danger of the ice melting.

- Since the sea levels are lifting up at an alarming rate of 3.2 *mm a year* there seem danger of the sea level rise around the world. Greenland region, has contributed in this context to about 0.5 *mm* till now.

- The estimation of the rise of 50 centimeter (20 inches), by 2100, with the lowest of 15 centimeters (6 inches) and the highest of 95 centimeters (37 inches), in the sea level due to the thermal expansion under the ocean and also because of the melting of the glaciers and ice sheets

- The mass of the ice located in Greenland is 680,000 m³, and it stretches up to 3 miles thick, covering three-quarters of the island. Some of the glaciers, particularly those in the southeast and northwest, have in the past 20 years dumped a increasing amount of ice into the ocean nearby, accounting for the rise of the water level by up to 15 % over the period.

- The planet is on a track to hit 2°C rise in the temperature if major steps to curb climate change aren't taken, and already much of the globe's warming has been absorbed by the oceans.

- Glacier has started a phase of self-sustained retreat and will irreversibly continue its decline. There may be a chance to grow glaciers at northern portion of Canada, USA, UK and may create venerable conditions of snow fall and cold waves in these regions and likely to force for shifting of living population at safer place.

- A calculation shows that between April 2003 and April 2012, the region was losing ice at the rate of 10 billion tons a year.

- By year 2100, if a minimum of **3.6 feet** (1.1 Meter) or maximum 10-13 feet (3.4-4 Meters) sea level rise occurs, then it will have a shift of ice melt into water by **397.245 trillion tonnes** or maximum 1100-1450 trillion tonnes respectively.

Thus, looking into the weight shift from polar (Northern / Southern coast) to sea, it might create change in the spinning angle of the Earth from 23.43 degree to further (+) or (-). The day may be a dark day on the beautiful planet when the entire living creatures may face dire consequences of their end up, provided things are checked and not to go beyond our control today. Try to imagine the consequences, act fast to "Save Earth; Save Life".

Acknowledgements

Enormous and extensive support provided by the management of School of Management Sciences, Lucknow, with a profound guidance under Harcourt Butler Technological Institute, Kanpur, has been a gliding way in accomplishing this project. Author Prof. Bharat Raj Singh expressed his special tribute to his mother Jagpatti Devi Singh who encouraged him to devote entire life for rendering the services to the human beings, though she had now left us for heavenly abode, leaving behind, her blessings for us.

Author details

Bharat Raj Singh[1*] and Onkar Singh[2]

*Address all correspondence to: brsinghlko@yahoo.com

1 School of Management Sciences, Technical Campus, Lucknow, India

2 Harcourt Butler Technological Institute, Kanpur, Uttar Pradesh, India

References

[1] Drinkwater, Mark; Kerr, Yann; Font, Jordi; Berger, Michael (February 2009). "Exploring the Water Cycle of the 'Blue Planet': The Soil Moisture and Ocean Salinity (SMOS) mission". *ESA Bulletin* (European Space Agency) (137): 6–15. "A view of Earth, the 'Blue Planet'... When astronauts first went into the space, they looked back at our Earth for the first time, and called our home the 'Blue Planet'."

[2] Dalrymple, G.B. (1991). *The Age of the Earth*. California: Stanford University Press. ISBN 0-8047-1569-6.

[3] Newman, William L. (2007-07-09). "Age of the Earth". Publications Services, USGS. Retrieved 2007-09-20.

[4] Dalrymple, G. Brent (2001). "The age of the Earth in the twentieth century: a problem (mostly) solved". *Geological Society, London, Special Publications* 190 (1): 205–221. Bibcode: 2001GSLSP.190.205D. doi:10.1144/GSL.SP.2001.190.01.14. Retrieved 2007-09-20.

[5] Stassen, Chris (2005-09-10). "The Age of the Earth". Talk Origins Archive. Retrieved 2008-12-30.

[6] Harrison, Roy M.; Hester, Ronald E. (2002). Causes and Environmental Implications of Increased UV-B Radiation. Royal Society of Chemistry, London; ISBN 0-85404-265-2.

[7] "NOAA – Ocean". Noaa.gov. Retrieved 2013-05-03.

[8] Yoder, Charles F. (1995). Astrometric and Geodetic Properties of Earth and the Solar System, T. J. Ahrens, ed. Global Earth Physics: A Handbook of Physical Constants.

Washington: American Geophysical Union. p. 8. ISBN 0-87590-851-9. Retrieved 2007-03-17.

[9] May, Robert M. (1988). "How many species are there on earth?". *Science* 241 (4872): 1441–1449. Bibcode: 1988Sci...241.1441M. doi:10.1126/science.241.4872.1441. PMID 17790039.

[10] United States Census Bureau (2 November 2011). "World POP Clock Projection". *United States Census Bureau International Database*. Retrieved 2011-11-02.

[11] Milbert, D. G.; Smith, D. A. (2007) "Converting GPS height into NAVD88 elevation with the GEOID96 geoid height model". National Geodetic Survey, NOAA. Retrieved 2007-03-07.

[12] Sandwell, D. T.; Smith, W. H. F. (2006-07-07). "Exploring the Ocean Basins with Satellite Altimeter Data". NOAA/NGDC. Retrieved 2007-04-21.

[13] Robert Krulwich, (2007),The 'Highest' Spot on Earth? NPR.org Consultado el 25-07-2010.

[14] Mohr, P. J.; Taylor, B. N. (October 2000). "Unit of length (meter)". *NIST Reference on Constants, Units, and Uncertainty*. NIST Physics Laboratory. Retrieved 2007-04-23.

[15] Staff (November 2001). "WPA Tournament Table & Equipment Specifications". World Pool-Billiards Association. Retrieved 2007-03-10.

[16] Senne, Joseph H. (2000). "Did Edmund Hillary Climb the Wrong Mountain". *Professional Surveyor* 20 (5): 16–21.

[17] Sharp, David (2005-03-05). "Chimborazo and the old kilogram". *The Lancet* 365 (9462): 831–832. doi:10.1016/S0140-6736(05)71021-7. PMID 15752514.

[18] Karl S. Kruszelnicki, (2004),"Tall Tales about Highest Peaks". Australian Broadcasting Corporation. Retrieved 2008-12-29.

[19] Willett, Sean D.; Brandon, Mark T. (January 2002). "On steady states in mountain belts". *Geology* 30 (2): 175–178. Bibcode: 2002Geo....30..175W. doi: 10.1130/0091-7613(2002)030<0175:OSSIMB>2.0.CO;2

[20] Summerfield, M.A., 1991, Global Geomorphology, Pearson Education Ltd, London. ISBN 0-582-30156-4, p 537.

[21] Dunai, T.J., 2010, Cosmogenic Nucleides, Cambridge University Press, 187 p. ISBN 978-0-521-87380-2.

[22] DTM intro page, Hunter College Department of Geography, New York NY, Website: http://www.geo.hunter.cuny.edu/terrain/intro.html

[23] Bennett, M. R., Glasser, N. F. (1996). Glacial Geology: Ice Sheets and Landforms, John Wiley & Sons Ltd, England. ISBN 0-471-96345-3.

[24] Church, Michael; Ryder, June M. (October 1972). "Paraglacial Sedimentation: A Consideration of Fluvial Processes Conditioned by Glaciation". *Geological Society of America Bulletin* 83 (10): 3059–3072. Bibcode: 1972GSAB...83.3059C. doi: 10.1130/0016-7606(1972)83[3059:PSACOF]2.0.CO;2

[25] Kargel, J. S., et al. (2012). Brief communication: Greenland's shrinking ice cover: 'Fast times' but not that fast. *The Cryosphere* 6, 533-537, doi: 10.5194/tc-6-533-2012.

[26] B.Wouters, J. L. Bamber, M. R. van den Broeke, J. T. M. Lenaerts and I. Sasgen, 2013, 'Limits in detecting acceleration of ice sheet mass loss due to climate variability', Nature Geoscience, July 16, 2013.

[27] Van den Broeke, M. *et al.* Partitioning recent greenland mass loss. Science 326, 984–986 (2009).

[28] Shfaqat A. Khan, et al., Sustained mass loss of the northeast Greenland ice sheet triggered by regional warming, Nature Climate Change, Vol. 4 (2014), pp 292–299, 16 March 2014, doi: 10.1038/nclimate2161

[29] Valentine, K. (2014). East Antarctic melting could raise sea levels by 10 To 13 feet, study finds, Journal Nature Climate Change, Florida, Article published on 5May 2014.

[30] Shepherd, A. *et al.* A reconciled estimate of ice-sheet mass balance. Science 338, 1183–1189 (2013).

[31] Kjær, K. H. *et al.* Aerial photographs reveal late-20th-century dynamic ice loss in northwestern Greenland. Science 337, 569–573 (2012).

[32] Moon, T., Joughin, I., Smith, B. & Howat, I. 21st-century evolution of Greenland outlet glacier velocities. Science 336, 576–578 (2012).

[33] Bjørk, A. A. *et al.* An aerial view of 80 years of climate-related glacier fluctuations in southeast Greenland. Nature Geosci. 5, 427–432 (2012).

[34] Khan, S. A. *et al.* Elastic uplift in southeast Greenland due to rapid ice mass loss. Geophys. Res. Lett. 34, L21701 (2007).

[35] Nick, F. M. *et al.* The response of Petermann Glacier, Greenland, to large calving events, and its future stability in the context of atmospheric and oceanic warming. J. Glaciol. Vol. 58 (208), p229-239, doi:10.3189/2012JoG11J242 (2012).

[36] Sasgen, I. *et al.* Timing and origin of recent regional ice-mass loss in Greenland. Earth Planet. Sc. Lett. 333–334, 29–303 (2012).

[37] Bamber, J. L. *et al.* A new bed elevation dataset for Greenland. Cryosphere 7, 499–510 (2013).

[38] Joughin, I. *et al.* Seasonal to decadal scale variations in the surface velocity of Jakobshavn Isbrae, Greenland: Observation and model-based analysis. J. Geophys. Res. 117, F02030 (2012).

[39] Krabill, W. B. *IceBridge ATM L2 Icessn Elevation, Slope, and Roughness, [1993–2012]. Boulder, Colorado, USA* (NASA Distributed Active Archive Center at the National Snow and Ice Data Center, http://nsidc.org/data/ilatm2.html (2012).

[40] GST *Ground control for 1:150,000 scale aerials, Greenland* (Danish Ministry of the Environment, Danish Geodata Agency, http://www.gst.dk/Emner/Referencenet/Referencesystemer/GR96/ (2013).

[41] Bevis, M. *et al.* Bedrock displacements in Greenland manifest ice mass variations, climate cycles and climate change. Proc. Natl Acad. Sci. USA 109, 11944–11948 (2012).

[42] Nick, F. *et al.* Future sea-level rise from Greenland's main outlet glaciers in a warming climate. Nature 497, 235–238 (2013).

[43] Price, S. F., Payne, A. J., Howat, I. M. & Smith, B. E Committed sea-level rise for the next century from Greenland ice sheet dynamics during the past decade. Proc. Natl Acad. Sci. USA 108, 8978–8983 (2011).

[44] Gillet-Chaulet, F. *et al.* Greenland ice sheet contribution to sea-level rise from a new-generation ice-sheet model. Cryosphere 6, 1561–1576 (2012).

[45] Yin, J. *et al.* Different magnitudes of projected subsurface ocean warming around Greenland and Antarctica. Nature Geosci. 4, 524–528 (2011).

[46] AMAP, *The Greenland Ice Sheet in a Changing Climate: Snow, Water, Ice and Permafrost in the Arctic (SWIPA)* (Arctic Monitoring and Assessment Programme, (2011).

[47] Zhang J., 2007: Increasing Antarctic Sea Ice under Warming Atmospheric and Oceanic Conditions, Journal of Climate, Vol. 20, 1 June 2007, pp 2515-2529.

[48] D. J. Cavalieri et al., 1997, Observed Hemispheric Asymmetry in Global Sea Ice Changes, *Science Magazine,* 7 November 1997,Vol. 278 no. 5340 pp. 1104-1106 DOI: 10.1126/science.278.5340.1104.

[49] Gloersen, P.; Campbell, W.J et al., 1992, Arctic and antarctic sea ice, 1978-1987: Satellite passive-microwave observations and analysis, NASA: Washington D.C. 290 pp.

[50] Streten, N. A., *et al.* (1980). Characteristics of the broadscale Antarctic sea ice extent and the associated atmospheric circulation 1972–1977. Australian Numerical Metrology research Centre, Melbourne, Volume 29, Issue 3, pp 279-299.

[51] Jain, S. (2014). Earth as a Planet, Fundamentals of Physical Geology, Springer Geology, New York, pp 57-75.

[52] Merritts, D., A. de Wet, and K. Menking, *Environmental Geology: an Earth System Science Approach.* New York, NY: W.H. Freeman and Company, 1998, chapter 1. ISBN: 9780716728344.

[53] Thompson, G. R., and J. Turk. *Environmental Geoscience.* 3rd ed. Ft Worth, TX: Harcourt Brace and Company, 1997. ISBN: 9780030988660.

PERMISSIONS

All chapters in this book were first published in GWCIR, by InTech Open; hereby published with permission under the Creative Commons Attribution License or equivalent. Every chapter published in this book has been scrutinized by our experts. Their significance has been extensively debated. The topics covered herein carry significant findings which will fuel the growth of the discipline. They may even be implemented as practical applications or may be referred to as a beginning point for another development.

The contributors of this book come from diverse backgrounds, making this book a truly international effort. This book will bring forth new frontiers with its revolutionizing research information and detailed analysis of the nascent developments around the world.

We would like to thank all the contributing authors for lending their expertise to make the book truly unique. They have played a crucial role in the development of this book. Without their invaluable contributions this book wouldn't have been possible. They have made vital efforts to compile up to date information on the varied aspects of this subject to make this book a valuable addition to the collection of many professionals and students.

This book was conceptualized with the vision of imparting up-to-date information and advanced data in this field. To ensure the same, a matchless editorial board was set up. Every individual on the board went through rigorous rounds of assessment to prove their worth. After which they invested a large part of their time researching and compiling the most relevant data for our readers.

The editorial board has been involved in producing this book since its inception. They have spent rigorous hours researching and exploring the diverse topics which have resulted in the successful publishing of this book. They have passed on their knowledge of decades through this book. To expedite this challenging task, the publisher supported the team at every step. A small team of assistant editors was also appointed to further simplify the editing procedure and attain best results for the readers.

Apart from the editorial board, the designing team has also invested a significant amount of their time in understanding the subject and creating the most relevant covers. They scrutinized every image to scout for the most suitable representation of the subject and create an appropriate cover for the book.

The publishing team has been an ardent support to the editorial, designing and production team. Their endless efforts to recruit the best for this project, has resulted in the accomplishment of this book. They are a veteran in the field of academics and their pool of knowledge is as vast as their experience in printing. Their expertise and guidance has proved useful at every step. Their uncompromising quality standards have made this book an exceptional effort. Their encouragement from time to time has been an inspiration for everyone.

The publisher and the editorial board hope that this book will prove to be a valuable piece of knowledge for researchers, students, practitioners and scholars across the globe.

LIST OF CONTRIBUTORS

Vytautas Pilipavičius
Aleksandras Stulginskis University, Faculty of Agronomy, Institute of Agroecosystems and Soil Sciences, Akademija, Lithuania

M. Dolores Garza-Gil, Manuel Varela-Lafuente, Gonzalo Caballero-Míguez and Julia Torralba-Cano
Department of Applied Economics, University of Vigo, Vigo, Spain

Akira Tomizuka
Graduate School of Fisheries Science and Environmental Studies, Nagasaki University, Japan

Akira Nishimura
Division of Mechanical Engineering, Graduate School of Engineering, Mie University, Tsu, Japan

Mohan Kolhe
Faculty of Engineering & Science, University of Agder, Grimstad, Norway

Sergio H. Franchito, J. P. R. Fernandez and David Pareja
Centro de Previsão de Tempo e Estudos Climáticos, CPTEC, Instituto Nacional de Pesquisas Espaciais, INPE, SP, Brazil

Georgios M. Photiadis
Centre for CO2 Technology, Department of Chemical Engineering, University College London, London, United Kingdom
Scientist in Raman Spectroscopy and in Molten Salt Chemistry and Technology, Potters Bar, Hertfordshire, England, United Kingdom.

Bharat Raj Singh
School of Management Sciences, Technical Campus, Lucknow, India

Onkar Singh
Harcourt Butler Technological Institute, Kanpur, Uttar Pradesh, India
Vice Chancellor, Madan Mohan Malviya University of Technology, Gorakhpur, Uttar Pradesh, India

Index

www.ingramcontent.com/pod-product-compliance
Lightning Source LLC
Chambersburg PA
CBHW082016190326
41458CB00010B/3206